CONTROL OF INDUCTION MOTORS

ACADEMIC PRESS SERIES IN ENGINEERING

Series Editor
J. David Irwin
Auburn University

This is a series that will include handbooks, textbooks, and professional reference books on cutting-edge areas of engineering. Also included in this series will be single-authored professional books on state-of-the-art techniques and methods in engineering. Its objective is to meet the needs of academic, industrial, and governmental engineers, as well as to provide instructional material for teaching at both the undergraduate and graduate level.

The series editor, J. David Irwin, is one of the best-known engineering educators in the world. Irwin has been chairman of the electrical engineering department at Auburn University for 27 years.

Published books in the series:
Embedded Microcontroller Interfacing for McoR Systems, 2000,
    G. J. Lipovski
Soft Computing & Intelligent Systems, 2000, N. K. Sinha, M. M. Gupta
Introduction to Microcontrollers, 1999, G. J. Lipovski
Industrial Controls and Manufacturing, 1999, E. Kamen
DSP Integrated Circuits, 1999, L. Wanhammar
Time Domain Electromagnetics, 1999, S. M. Rao
Single- and Multi-Chip Microcontroller Interfacing, 1999, G. J. Lipovski
Control in Robotics and Automation, 1999, B. K. Ghosh, N. Xi,
    and T. J. Tarn

CONTROL OF INDUCTION MOTORS

Andrzej M. Trzynadlowski
Department of Electrical Engineering
University of Nevada, Reno
Reno, Nevada

ACADEMIC PRESS

A Harcourt Science and Technology Company

San Diego San Francisco New York Boston
London Sydney Tokyo

ACADEMIC PRESS
A Harcourt Science and Technology Company
525 B Street, Suite 1900, San Diego, CA 92101-4495, USA
http://www.academicpress.com

Academic Press
Harcourt Place, 32 Jamestown Road, London NW1 7BY, UK

Library of Congress Catalog Card Number: 00-104379
International Standard Book Number: 0-12-701510-8

Transferred to Digital Printing 2008
ISBN: 9780127015101

To my wife, Dorota, and children, Bart and Nicole

CONTENTS

Preface   xi

1 Background   1

1.1   Induction Motors   1
1.2   Drive Systems with Induction Motors   3
1.3   Common Loads   4
1.4   Operating Quadrants   10
1.5   Scalar and Vector Control Methods   11
1.6   Summary   12

2 Construction and Steady-State Operation of Induction Motors   15

2.1   Construction   15
2.2   Revolving Magnetic Field   17
2.3   Steady-State Equivalent Circuit   24
2.4   Developed Torque   27

2.5   Steady-State Characteristics   31
2.6   Induction Generator   36
2.7   Summary   40

3 Uncontrolled Induction Motor Drives   43

3.1   Uncontrolled Operation of Induction Motors   43
3.2   Assisted Starting   44
3.3   Braking and Reversing   47
3.4   Pole Changing   51
3.5   Abnormal Operating Conditions   52
3.6   Summary   54

4 Power Electronic Converters for Induction Motor Drives   55

4.1   Control of Stator Voltage   55
4.2   Rectifiers   56
4.3   Inverters   64
4.4   Frequency Changers   69
4.5   Control of Voltage Source Inverters   71
4.6   Control of Current Source Inverters   81
4.7   Side Effects of Converter Operation in Adjustable Speed Drives   88
4.8   Summary   91

5 Scalar Control Methods   93

5.1   Two-Inductance Equivalent Circuits of the Induction Motor   93
5.2   Open-Loop Scalar Speed Control (Constant Volts/Hertz)   97
5.3   Closed-Loop Scalar Speed Control   101
5.4   Scalar Torque Control   102
5.5   Summary   105

6 Dynamic Model of the Induction Motor   107

6.1   Space Vectors of Motor Variables   107
6.2   Dynamic Equations of the Induction Motor   111
6.3   Revolving Reference Frame   114
6.4   Summary   117

7 Field Orientation   119

7.1   Torque Production and Control in the DC Motor   119
7.2   Principles of Field Orientation   121
7.3   Direct Field Orientation   124
7.4   Indirect Field Orientation   126
7.5   Stator and Airgap Flux Orientation   129
7.6   Drives with Current Source Inverters   134
7.7   Summary   135

8 Direct Torque and Flux Control   137

8.1   Induction Motor Control by Selection of Inverter States   137
8.2   Direct Torque Control   140
8.3   Direct Self-Control   148
8.4   Space-Vector Direct Torque and Flux Control   145
8.5   Summary   157

9 Speed and Position Control   159

9.1   Variables Controlled in Induction Motor Drives   159
9.2   Speed Control   161
9.3   Machine Intelligence Controllers   164
9.4   Position Control   173
9.5   Summary   175

10 Sensorless Drives   177

10.1 Issues in Sensorless Control of Induction Motors   177
10.2 Flux Calculators   179
10.3 Speed Calculators   183
10.4 Parameter Adaptation and Self-Commissioning   191
10.5 Commercial Adjustable Speed Drives   197
10.6 Summary   200

Literature   203
Glossary of Symbols   221
Index   225

PREFACE

More than half of the total electrical energy produced in developed countries is converted into mechanical energy in electric motors, freeing the society from the tedious burden of physical labor. Among many types of the motors, three-phase induction machines still enjoy the same unparalleled popularity as they did a century ago. At least 90% of industrial drive systems employ induction motors.

Most of the motors are uncontrolled, but the share of adjustable speed induction motor drives fed from power electronic converters is steadily increasing, phasing out dc drives. It is estimated that more than 50 billion dollars could be saved annually by replacing all "dumb" motors with controlled ones. However, control of induction machines is a much more challenging task than control of dc motors. Two major difficulties are the necessity of providing adjustable-frequency voltage (dc motors are controlled by adjusting the *magnitude* of supply voltage) and the nonlinearity and complexity of analytical model of the motor, aggrandized by parameter uncertainty.

As indicated by the title, this book is devoted to various aspects of control of induction motors. In contrast to the several existing monographs on adjustable speed drives, a great effort was made to make the covered topics easy to understand by nonspecialists. Although primarily addressed to practicing engineers, the book may well be used as a graduate textbook or an auxiliary reference source in undergraduate courses on electrical machinery, power electronics, or electric drives.

Beginning with a general background, the book describes construction and steady-state operation of induction motors and outlines basic issues in uncontrolled drives. Power electronic converters, especially pulse width modulated inverters, constitute an important part of adjustable speed drives. Therefore, a whole chapter has been devoted to them. The part of the book dealing with control topics begins with scalar control methods used in low-performance drive systems. The dynamic model of the induction machine is introduced next, as a base for presentation of more advanced control concepts. Principles of the field orientation, a fundamental idea behind high-performance, vector controlled drives, are then elucidated. The book also shows in detail another common approach to induction motor control, the direct torque and flux control, and use of induction motors in speed and position control systems is illustrated. Finally, the important topic of sensorless control is covered, including a brief review of the commercial drives available on today's market.

Certain topics encountered in the literature on induction motor drives have been left out. The issue of control of this machine is so intellectually challenging that some researchers attempt approaches fundamentally different from the established methods. As of now, such ideas as feedback linearization or passivity based control have not yet found their way to practical ASDs. Time will show whether these theoretical concepts represent a sufficient degree of improvement over the existing techniques to enter the domain of commercial drives.

Selected literature, a glossary of symbols, and an index complete the book. Easy-to-follow examples illustrate the presented ideas. Numerous figures facilitate understanding of the text. Each chapter begins with a short abstract and ends with a summary, following the three tenets of good teaching philosophy: (1) Tell what you are going to tell, (2) tell, and (3) then tell what you just told.

I want to thank Professor J. David Irwin of Auburn University for the encouragement to undertake this serious writing endeavor. My wife, Dorota, and children, Bart and Nicole, receive my deep gratitude for their sustained support.

1

BACKGROUND

In this introductory chapter, a general characterization of induction motors and their use in ac drive systems is given. Common mechanical loads and their characteristics are presented, and the concept of operating quadrants is explained. Control methods for induction motors are briefly reviewed.

1.1 INDUCTION MOTORS

Three-phase induction motors are so common in industry that in many plants no other type of electric machine can be found. The author remembers his conversation with a maintenance supervisor in a manufacturing facility who, when asked what types of motors they had on the factory floor, replied: "Electric motors, of course. What else?" As it turned out, all the motors, hundreds of them, were of the induction, *squirrel-cage* type. This simple and robust machine, an ingenious invention of the late nineteenth century, still maintains its unmatched popularity in industrial practice.

Induction motors employ a simple but clever scheme of electromechanical energy conversion. In the squirrel-cage motors, which constitute a vast majority of induction machines, the rotor is inaccessible. No moving contacts, such as the commutator and brushes in dc machines or slip rings and brushes in ac synchronous motors and generators, are needed. This arrangement greatly increases reliability of induction motors and eliminates the danger of sparking, permitting squirrel-cage machines to be safely used in harsh environments, even in an explosive atmosphere. An additional degree of ruggedness is provided by the lack of wiring in the rotor, whose winding consists of uninsulated metal bars forming the "squirrel cage" that gives the name to the motor. Such a robust rotor can run at high speeds and withstand heavy mechanical and electrical overloads. In adjustable-speed drives (ASDs), the low electric time constant speeds up the dynamic response to control commands. Typically, induction motors have a significant torque reserve and a low dependence of speed on the load torque.

The less common *wound-rotor* induction motors are used in special applications, in which the existence and accessibility of the rotor winding is an advantage. The winding can be reached via brushes on the stator that ride atop slip rings on the rotor. In the simplest case, adjustable resistors (rheostats) are connected to the winding during startup of the drive system to reduce the motor current. Terminals of the winding are shorted when the motor has reached the operating speed. In the more complicated so-called *cascade systems*, excess electric power is drawn from the rotor, conditioned, and returned to the supply line, allowing speed control. A price is paid for the extra possibilities offered by wound-rotor motors, as they are more expensive and less reliable than their squirrel-cage counterparts. In today's industry, wound-rotor motors are increasingly rare, having been phased out by controlled drives with squirrel-cage motors. Therefore, only the latter motors will be considered in this book.

Although operating principles of induction motors have remained unchanged, significant technological progress has been made over the years, particularly in the last few decades. In comparison with their ancestors, today's motors are smaller, lighter, more reliable, and more efficient. The so-called *high-efficiency motors*, in which reduced-resistance windings and low-loss ferromagnetic materials result in tangible savings of consumed energy, are widely available. High-efficiency motors are somewhat more expensive than standard machines, but in most applications the simple payback period is short. Conservatively, the average life span of an induction motor can be assumed to be about 12 years (although properly maintained motors can work for decades), so replacement of a worn standard motor with a high-efficiency one that would pay off for its higher price in, for instance, 2 years, is a simple matter of common sense.

1.2 DRIVE SYSTEMS WITH INDUCTION MOTORS

An electric motor driving a mechanical load, directly or through a gearbox or a V-belt transmission, and the associated control equipment such as power converters, switches, relays, sensors, and microprocessors, constitute an *electric drive system*. It should be stressed that, as of today, most induction motor drives are still basically uncontrolled, the control functions limited to switching the motor on and off. Occasionally, in drive systems with difficult start-up due to a high torque and/or inertia of the load, simple means for reducing the starting current are employed. In applications where the speed, position, or torque must be controlled, ASDs with dc motors are still common. However, ASDs with induction motors have increasing popularity in industrial practice. The progress in control means and methods for these motors, particularly spectacular in the last decade, has resulted in development of several classes of ac ASDs having a clear competitive edge over dc drives.

Most of the energy consumed in industry by induction motors can be traced to high-powered but relatively unsophisticated machinery such as pumps, fans, blowers, grinders, or compressors. Clearly, there is no need for high dynamic performance of these drives, but speed control can bring significant energy savings in most cases. Consider, for example, a constant-speed blower, whose output is regulated by choking the air flow in a valve. The same valve could be kept fully open at all times (or even disposed of) if the blower were part of an adjustable-speed drive system. At a low air output, the motor would consume less power than that in the uncontrolled case, thanks to the reduced speed and torque.

High-performance induction motor drives, such as those for machine tools or elevators, in which the precise torque and position control is a must, are still relatively rare, although many sophisticated control techniques have already reached the stage of practicality. For better driveability, high-performance adjustable-speed drives are also increasingly used in electrical traction and other electric vehicles.

Except for simple two-, three-, or four-speed schemes based on *pole changing*, an induction motor ASD must include a variable-frequency source, the so-called *inverter*. Inverters are dc to ac converters, for which the dc power must be supplied by a *rectifier* fed from the ac power line. The so-called *dc link*, in the form of a capacitor or reactor placed between the rectifier and inverter, gives the rectifier properties of a voltage source or a current source. Because rectifiers draw distorted, nonsinusoidal currents from the power system, passive or active filters are required at their input to reduce the low-frequency harmonic content in the supply currents. Inverters, on the other hand, generate high-frequency current noise, which must not be allowed to reach the system. Otherwise, operation of sensitive communication and control equipment could be disturbed by the resultant *electromagnetic interference* (EMI). Thus, effective EMI filters are needed too.

For control of ASDs, microcomputers, microcontrollers, and digital signal processors (DSPs) are widely used. When sensors of voltage, current, speed, or position are added, an ASD represents a much more complex and expensive proposition than does an uncontrolled motor. This is one reason why plant managers are so often wary of installing ASDs. On the other hand, the motion-control industry has been developing increasingly efficient, reliable, and user-friendly systems, and in the time to come ASDs with induction motors will certainly gain a substantial share of industrial applications.

1.3 COMMON LOADS

Selection of an induction motor and its control scheme depends on the load. An ASD of a fan will certainly differ from that of a winder in a paper mill, the manufacturing process in the latter case imposing narrow tolerance bands on speed and torque of the motor. Various classifications can be used with respect to loads. In particular, they can be classified with respect to: (a) inertia, (b) torque versus speed characteristic, and (c) control requirements.

High-inertia loads, such as electric vehicles, winders, or centrifuges, are more difficult to accelerate and decelerate than, for instance, a pump or a grinder. The total mass moment of inertia referred to the motor shaft can be computed from the kinetic energy of the drive. Consider, for example, a motor with the rotor inertia of J_M that drives a load with the mass moment of inertia of J_L through a transmission with the gear ratio of N. The kinetic energy, E_L, of the load rotating with the angular velocity ω_L is

$$E_L = \frac{J_L \omega_L^2}{2}, \tag{1.1}$$

while the kinetic energy, E_M, of the rotor whose velocity is ω_M is given by

$$E_M = \frac{J_M \omega_M^2}{2}. \tag{1.2}$$

Thus, the total kinetic energy, E_T, of the drive can be expressed as

$$E_T = E_L + E_M = \left[\left(\frac{\omega_L}{\omega_M} \right)^2 J_L + J_M \right] \frac{\omega_M^2}{2} = \frac{J_T \omega_M^2}{2}, \tag{1.3}$$

where J_T denotes the total mass moment of inertia of the system referred to the motor shaft. Because

$$\frac{\omega_L}{\omega_M} = N, \tag{1.4}$$

then

$$J_T = N^2 J_L + J_M. \tag{1.5}$$

The difference, T_d, between the torque, T_M, developed in the motor and the *static torque*, T_L, with which the load resists the motion is called a *dynamic torque*. According to Newton's second law,

$$T_d = T_M - T_L = J_T \frac{d\omega_M}{dt} = \frac{J_T}{N} \frac{d\omega_L}{dt}, \tag{1.6}$$

or

$$\frac{d\omega_L}{dt} = \frac{N T_d}{J_T}. \tag{1.7}$$

Unsurprisingly, the preceding equation indicates that a high mass moment of inertia makes a drive sluggish, so that a high dynamic torque is required for fast acceleration or deceleration of the load.

The concept of an *equivalent wheel* is convenient for calculation of the total mass moment of inertia referred to the shaft of a motor driving an electric vehicle or another linear-motion load. The equivalent wheel is a hypothetical wheel assumed to be directly driven by the motor and whose peripheral velocity, u_L, equals the linear speed of the load. Denoting the radius of the equivalent wheel by r_{eq}, the speed of the load can be expressed in terms of that radius and motor speed as

$$u_L = r_{eq} \omega_M. \tag{1.8}$$

The equation for kinetic energy, E_L, of the load whose mass is denoted by m_L,

$$E_L = \frac{m_L u_L^2}{2},\tag{1.9}$$

can therefore be rearranged to

$$E_L = \frac{m_L (r_{eq}\omega_M)^2}{2} = \frac{J_L \omega_M^2}{2},\tag{1.10}$$

with J_L denoting the effective mass moment of inertia of the load, given by

$$J_L = m_L r_{eq}^2.\tag{1.11}$$

Because the motor is assumed to drive the equivalent wheel directly (i.e., $N = 1$), the total mass moment of inertia, J_T, is equal to the sum of J_L and J_M.

EXAMPLE 1.1 If a vehicle is driven by several motors (as, for instance, in an electric locomotive) the load inertia seen by a single motor represents a respective fraction of J_L. Determine the load mass moment of inertia per single motor of a freight train hauled by three locomotives, each driven by ten motors. The train weighs 20,000 tons, and, when it runs at 50 mph, rotors of the motors rotate at 1500 rpm.

The equivalent-wheel radius equals (50 × 1609 m / 3600 s) / (1500 × 2π rad / 60 s) = 0.142 m. According to Eq. (1.11), the total mass moment of inertia of the load equals 20,000 × 1000 kg × (0.142 m)² = 403,280 kg.m². The fractional mass moment of inertia seen by each of the 30 motors equals 403,280 kg.m² / 30 = 13,443 kg.m², still an enormous value. ∎

In most loads, the static torque, T_L, depends on the load speed, ω_L. The $T_L(\omega_L)$ relation, usually called a mechanical characteristic, is an important feature of the load, because its intersection with the analogous characteristic of the motor, $T_M(\omega_M)$, determines the steady-state operating point of the drive. Expressing the mechanical characteristic by a general equation

$$T_L = T_{L0} + \tau \omega_L^k,\tag{1.12}$$

where T_{L0} and τ are constants, three basic types, illustrated in Figure 1.1, can be distinguished:

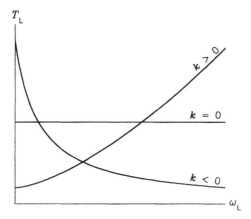

FIGURE 1.1   Mechanical characteristics of common loads.

1. *Constant-torque* characteristic, with k $\approx$ 0, typical for lifts and conveyors and, generally, for loads whose speed varies in a narrow range only.
2. *Progressive-torque* characteristic, with $k > 0$, typical for pumps, fans, blowers, compressors, electric vehicles and, generally, for most loads with a widely varying speed.
3. *Regressive-torque* characteristic, with $k < 0$, typical for winders. There, with a constant tension and linear speed of the wound tape, an increase in the coil radius is accompanied by a decreasing speed and an increasing torque.

Practical loads are better described by *operating areas* rather than mechanical characteristics. An operating area represents a set of all allowable operating points in the (ω_L, T_L) plane. Taking a pump as an example, its torque versus speed characteristic strongly depends on the pressure and viscosity of the pumped fluid. Analogously, the mechanical characteristic of a winder varies with changes in the tape tension and speed. Therefore, a single mechanical characteristic cannot account for all possible operating points. An example operating area of a progressive-torque load is shown in Figure 1.2a. Clearly, if a load is driven directly by a motor, the motor operating area in the (ω_M, T_M) plane is the same as that of the load. However, if the load is geared to the motor, the operating areas of the load and motor differ because the gearing acts as a transformer of the mechanical power. The operating area of a motor driving the load in Figure 1.2a through a frictionless transmission with a gear ratio of 0.5 is shown in Figure 1.2b.

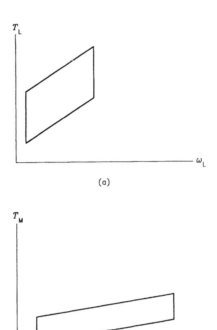

FIGURE 1.2   Example operating areas: (a) load, (b) motor (same speed and torque scales used in both diagrams).

EXAMPLE 1.2   The coil radius, r, in a textile winder changes from 0.15 m (empty coil) to 0.5 m (full coil). The automatically controlled tension, F, of the wound fabric can be set to any value between 100 N and 500 N, and the linear speed, u, of the fabric is adjustable within the 2 m/s to 4.8 m/s range. Determine the operating area of the winder.

The constant-force, constant-speed operation of the winder makes the exponent k in Eq. (1.12) equal to -1. Indeed, because

$$\omega_L = \frac{u}{r},$$

and

$$T_L = Fr,$$

then

$$T_L = \frac{Fu}{\omega_L}.$$

Assuming that the tension and speed of the fabric can be set to any allowable value, independently from each other, the operating speed of the winder is limited to the $1/0.5 = 2$ rad/s to $2.4/0.15 = 16$ rad/s range. If expressed in r/min, this speed range is 19.1 r/min to 152.8 r/min. The operating area, shown in Figure 1.3, is bound by two hyperbolic curves corresponding to the minimum and maximum values of force and speed. ■

In a properly designed drive system, the motor operates safely at every point of its operating area, that is, neither the voltage, current, nor speed exceeds its allowable values. The gearing may be needed to provide proper matching of the motor to the load. A gear ratio less than unity is employed when the load is to run slower than the motor, with a torque greater than that of the motor. Conversely, a high-speed, low-torque load requires a gear ratio greater than unity.

Control requirements depend on the particular application of a drive system. In most practical drives, such as those of pumps, fans, blowers, conveyors, or centrifuges, the main controlled variable is the load speed. High control accuracy in such systems is usually not necessary. Drives with a directly controlled torque, for instance those of winders or electric vehicles, are more demanding with regard to the control quality. Finally, positioning systems, such as precision machine tools or elevator drives, must be endowed with the highest level of dynamic performance. In certain positioning systems, control requirements are so strict that induction motors cannot be employed.

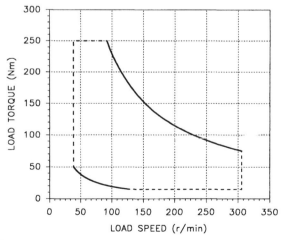

FIGURE 1.3 Operating area of the example winder.

1.4 OPERATING QUADRANTS

The concept of *operating quadrants* plays an important role in the theory and practice of electric drives. Both the torque, T_M, developed in a motor and speed, ω_M, of the rotor can assume two polarities. For instance, watching the motor from the front end, positive polarity can be assigned to the clockwise direction and negative polarity to the counterclockwise direction. Because the output (mechanical) power, P_{out}, of a motor is given by

$$P_{out} = T_M \omega_M, \tag{1.13}$$

the torque and speed polarities determine the direction of flow of power between the motor and load. With $P_{out} > 0$, the motor draws electric power from a supply system and converts it into mechanical power delivered to the load. Conversely, $P_{out} < 0$ indicates a reversed power flow, with the motor being driven by the load that acts as a *prime mover*. If proper arrangements are made, the motor can then operate as a generator and deliver electric power to the supply system. Such a regenerative mode of operation can be employed for braking a high-inertia load or lowering a load in a lift drive, reducing the net energy consumption by the motor.

The operating quadrants in the already mentioned (ω_M, T_M) plane correspond to the four possible combinations of polarities of torque and speed, as shown in Figure 1.4. The power flow in the first quadrant and third quadrant is positive, and it is negative in the second and fourth quadrants. To illustrate the idea of operating quadrants, let us consider two drive systems, that of an elevator and that of an electric locomotive. When lifting, the torque and speed of elevator's motor have the same

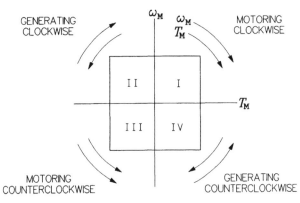

FIGURE 1.4   Operating quadrants in the (ω_M, T_M) plane.

polarity. However, when lowering, the motor rotates in the other direction while the polarity of the torque remains unchanged. Indeed, in both cases the motor torque must counterbalance the unidirectional gravity torque. Thus, assuming a positive motor speed when lifting, the motor is seen to operate in the first quadrant, while operation in the fourth quadrant occurs when lowering. In the latter situation, it is the weight of the elevator cage that drives the motor, and the potential energy of the cage is converted into electrical energy in the motor. The supply system of the motor must be so designed that this energy is safely dissipated or returned to the power source.

As for the locomotive, both polarities of the motor speed are possible, depending on the direction of linear motion of the vehicle. Also, the motor torque can assume two polarities, agreeing with the speed when the locomotive is in the driving mode and opposing the speed when braking. The enormous kinetic energy would strain the mechanical brakes if they were the only source of braking torque. Therefore, all electric locomotives (and other electric vehicles as well) have a provision allowing *electrical braking*, which is performed by forcing the motor to operate as a generator. It can be seen that the two possible polarities of both the torque and speed make up for four quadrants of operation of the drive. For example, first quadrant may correspond to the forward driving, second quadrant to the forward braking, third quadrant to the backward driving, and fourth quadrant to the backward braking. Yet, it is worth mentioning that, apart from electric vehicles, the four-quadrant operation is not very common in practice. Most of the ASDs, as well as uncontrolled motors, operate in the first quadrant only.

Power electronic converters feeding induction motors in ASDs also can operate in up to four quadrants in the current-voltage plane. As known from the theory of electric machines, the developed torque and the armature current are closely related. The same applies to the speed and armature voltage of a machine. Therefore, if a converter-fed motor operates in a certain quadrant, the converter operates in the same quadrant.

1.5 SCALAR AND VECTOR CONTROL METHODS

Induction motors can be controlled in many ways. The simplest methods are based on changing the structure of stator winding. Using the so-called *wye-delta switch*, the starting current can easily be reduced. Another type of switch allows emulation of a gear change by the already-mentioned pole changing, that is, changing the number of magnetic poles of the stator. However, in modern ASDs, it is the stator voltage and current that are subject to control. These, in the steady state, are defined by their magnitude and frequency; and if these are the parameters that are adjusted, the control technique belongs in the class of *scalar control* methods. A rapid change in the magnitude or frequency may produce undesirable transient effects, for example a disturbance of the normally constant motor torque. This, fortunately, is not important in low-performance ASDs, such as those of pumps, fans, or blowers. There, typically, the motor speed is open-loop controlled, with no speed sensor required (although current sensors are usually employed in overcurrent protection circuits).

In high-performance drive systems, in which control variables include the torque developed in the motor, *vector control* methods are necessary. The concept of space vectors of motor quantities will be explained later. Here, it is enough to say that a vector represents *instantaneous* values of the corresponding three-phase variables. For instance, the vector of stator current is obtained from the currents in all three phases of the stator and, conversely, all three phase currents can be determined from the current vector. In vector control schemes, space vectors of three-phase motor variables are manipulated according to the control algorithm. Such an approach is primarily designed for maintaining continuity of the torque control during transient states of the drive system.

Needless to say, vector control systems are more complex than those realizing the scalar control. Voltage and current sensors are always used; and, for the highest level of performance of the ASD, speed and position sensors may be necessary as well. Today, practically all control systems for electric motors are based on digital integrated circuits of some kind, such as microcomputers, microcontrollers, or digital signal processors (DSPs).

1.6 SUMMARY

Induction motors, especially those of the squirrel-cage type, are the most common sources of mechanical power in industry. Supplied from a three-phase ac line, they are simple, robust, and inexpensive. Although most motors operate with a fixed frequency resulting in an almost constant speed, ASDs are increasingly introduced in a variety of applications. Such a drive must include a power electronic converter to control the magnitude and frequency of the voltage and current supplied to the motor. A control system governing the operation of the drive system is usually of the digital type.

Common mechanical loads can be classified with respect to their inertia, to the torque-speed characteristic (mechanical characteristic), and to the control requirements. Depending on the particular application, the driving motor may operate in a single quadrant, two quadrants, or four quadrants of the (ω_M, T_M) plane.

Scalar control methods, in which only the magnitude and frequency of the fundamental voltage and current supplied to the motor are adjusted, are employed in low-performance drives. If high dynamic performance of a drive is required under both the steady-state and transient operating conditions, vector techniques are used to adjust the instantaneous values of voltage and current.

2

CONSTRUCTION AND STEADY-STATE OPERATION OF INDUCTION MOTORS

Construction and operating principles of induction motors are presented in this chapter. The generation of a revolving magnetic field in the stator and torque production in the rotor are described. The per-phase equivalent circuit is introduced for determination of steady-state characteristics of the motor. Operation of the induction machine as a generator is explained.

2.1 CONSTRUCTION

An induction motor consists of many parts, the stator and rotor being the basic subsystems of the machine. An exploded view of a squirrel-cage motor is shown in Figure 2.1. The motor case (frame), ribbed outside for better cooling, houses the stator core with a three-phase winding placed in slots on the periphery of the core. The stator core is made of thin (0.3 mm to 0.5 mm) soft-iron laminations, which are stacked and screwed together. Individual laminations are covered on both sides with insulating lacquer to reduce eddy-current losses. On the front side, the stator housing is closed by a cover, which also serves as a support for the front bearing

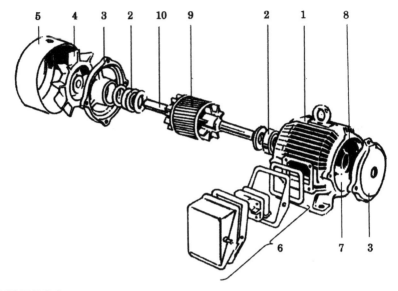

FIGURE 2.1 Exploded view of an induction motor: (1) motor case (frame), (2) ball bearings, (3) bearing holders, (4) cooling fan, (5) fan housing, (6) connection box, (7) stator core, (8) stator winding (not visible), (9) rotor, (10) rotor shaft. *Courtesy of Danfoss A/S.*

of the rotor. Usually, the cover has drip-proof air intakes to improve cooling. The rotor, whose core is also made of laminations, is built around a shaft, which transmits the mechanical power to the load. The rotor is equipped with cooling fins. At the back, there is another bearing and a cooling fan affixed to the rotor. The fan is enclosed by a fan cover. Access to the stator winding is provided by stator terminals located in the connection box that covers an opening in the stator housing.

Open-frame, partly enclosed, and *totally enclosed* motors are distinguished by how well the inside of stator is sealed from the ambient air. Totally enclosed motors can work in extremely harsh environments and in explosive atmospheres, for instance, in deep mines or lumber mills. However, the cooling effectiveness suffers when the motor is tightly sealed, which reduces its power rating.

The squirrel-cage rotor winding, illustrated in Figure 2.2, consists of several bars connected at both ends by end rings. The rotor cage shown is somewhat oversimplified, practical rotor windings being made up of more than few bars (e.g., 23), not necessarily round, and slightly skewed with respect to the longitudinal axis of the motor. In certain machines, in order to change the inductance-to-resistance ratio that strongly influences mechanical characteristics of the motor, rotors with deep-bar cages and

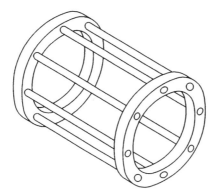

FIGURE 2.2   Squirrel-cage rotor winding.

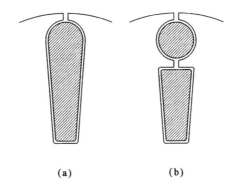

(a)   (b)

FIGURE 2.3   Cross-section of a rotor bar in (a) deep-bar cage, (b) double cage.

double cages are used. Those are depicted in Figures 2.3a and 2.3b, respectively.

2.2 REVOLVING MAGNETIC FIELD

The three-phase stator winding produces a revolving magnetic field, which constitutes an important property of not only induction motors but also synchronous machines. Generation of the revolving magnetic field by stationary phase windings of the stator is explained in Figures 2.4 through 2.9. A simplified arrangement of the windings, each consisting of a one-loop single-wire coil, is depicted in Figure 2.4 (in real motors, several multiwire loops of each phase winding are placed in slots spread along the inner periphery of the stator). The coils are displaced *in space* by 120° from each other. They can be connected in wye or delta, which in

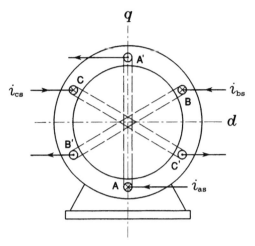

FIGURE 2.4 Two-pole stator of the induction motor.

this context is unimportant. Figure 2.5 shows waveforms of currents i_{as}, i_{bs}, and i_{cs} in individual phase windings. The stator currents are given by

$$i_{as} = I_{s,m}\cos(\omega t), \tag{2.1}$$

$$i_{bs} = I_{s,m}\cos\left(\omega t - \frac{2}{3}\pi\right),$$

and

$$i_{cs} = I_{s,m}\cos\left(\omega t - \frac{4}{3}\pi\right),$$

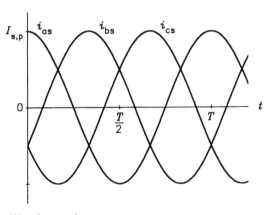

FIGURE 2.5 Waveforms of stator currents.

where $I_{\mathrm{s,p}}$ denotes their peak value and ω is the supply radian frequency; they are mutually displaced *in phase* by the same 120°. A phasor diagram of stator currents, at the instant of $t = 0$, is shown in Figure 2.6 with the corresponding distribution of currents in the stator winding. Current entering a given coil at the end designated by an unprimed letter, e.g., A, is considered positive and marked by a cross, while current leaving a coil at that end is marked by a dot and considered negative. Also shown are vectors of the magnetomotive forces (MMFs), $\mathscr{F}_{\mathrm{sa}}$, $\mathscr{F}_{\mathrm{sb}}$, and $\mathscr{F}_{\mathrm{sc}}$, produced by the phase currents. These, when added, yield the vector, $\mathscr{F}_{\mathrm{s}}$, of the total MMF of the stator, whose magnitude is 1.5 times greater than that of the maximum value of phase MMFs. The two half-circular loops represent the pattern of the resultant magnetic field, that is, lines of the magnetic flux, ϕ_{s}, of stator.

At $t = T/6$, where T denotes the period of stator voltage, that is, a reciprocal of the supply frequency, f, the phasor diagram and distribution

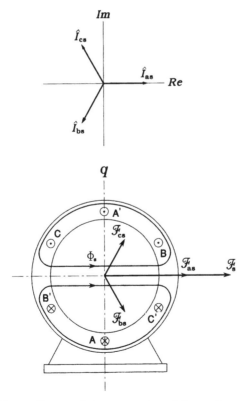

FIGURE 2.6 Phasor diagram of stator currents and the resultant magnetic field in a two-pole motor at $\omega t = 0$.

of phase currents and MMFs are as seen in Figure 2.7. The voltage phasors have turned counterclockwise by 60°. Although phase MMFs did not change their directions, remaining perpendicular to the corresponding stator coils, the total MMF has turned by the same 60°. In other words, the spacial angular displacement, α, of the stator MMF equals the "electric angle," ωt. In general, production of a revolving field requires at least two phase windings displaced *in space*, with currents in these windings displaced *in phase*.

The stator in Figure 2.4 is called a two-pole stator because the magnetic field, which is generated by the total MMF and which closes through the iron of the stator and rotor, acquires the same shape as that produced by two revolving physical magnetic poles. A four-pole stator is shown in Figure 2.8 with the same values of phase currents as those in Figure 2.6. When, $T/6$ seconds later, the phasor diagram has again turned by 60°, the pattern of crosses and dots marking currents in individual conductors of

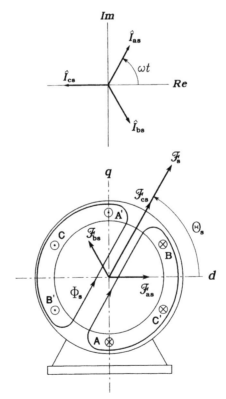

FIGURE 2.7 Phasor diagram of stator currents and the resultant magnetic field in a two-pole motor at $\omega t = 60°$.

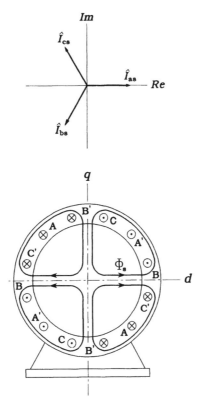

FIGURE 2.8   Phasor diagram of stator currents and the resultant magnetic field in a four-pole motor at $\omega t = 0$.

the stator has turned by 30° only, as seen in Figure 2.9. Clearly, the total MMF has turned by the same spacial angle, α, which is now equal to a half of the electric angle, ωt. The magnetic field is now as if it were generated by four magnetic poles, N-S-N-S, displaced by 90° from each other on the inner periphery of the stator. In general,

$$\alpha = \frac{\omega t}{p_p}, \tag{2.2}$$

where p_p denotes the number of pole pairs. Dividing both sides of Eq. (2.2) by t, the angular velocity, ω_{syn}, of the field, called a *synchronous velocity*, is obtained as

$$\omega_{syn} = \frac{\omega}{p_p}, \tag{2.3}$$

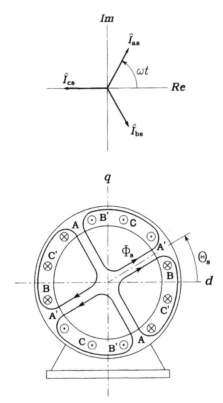

FIGURE 2.9 Phasor diagram of stator currents and the resultant magnetic field in a four-pole motor at $\omega t = 60°$.

while the *synchronous speed*, n_{syn}, of the field in revolutions per minute (r/min) is

$$n_{\text{syn}} = \frac{60}{p_{\text{p}}} f. \tag{2.4}$$

To explain how a torque is developed in the rotor, consider an arrangement depicted in Figure 2.10 and representing an "unfolded" motor. Conductor CND, a part of the squirrel-cage rotor winding, moves leftward with the speed u_1. The conductor is immersed in a magnetic field produced by stator winding and moving leftward with the speed u_2, which is greater than u_1. The field is marked by small crossed circles representing lines of magnetic flux, ϕ, directed toward the page. Thus, with respect to the field, the conductor moves to the right with the speed $u_3 = u_2 - u_1$. This motion induces (hence the name of the motor) an electromotive force (EMF), e, whose polarity is determined by the well-known right-hand

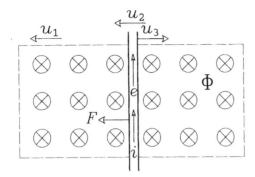

FIGURE 2.10   Generation of electrodynamic force in a rotor bar of the induction motor.

rule. Clearly, no EMF would be induced if the speed of the conductor (i.e., that of the rotor) and speed of the field were equal, because according to Faraday's law the EMF is proportional to the rate of change of flux linkage of the conductor. If the conductor was stationary with respect to the field, that is, if the rotor rotated with the synchronous speed, no changes would be experienced in the flux linking the conductor.

The EMF, e, produces a current, i, in the conductor. The interaction of the current and magnetic field results in an electrodynamic force, F, generated in the conductor. The left-hand rule determines direction of the force. It is seen that the force acts on the conductor in the same direction as that of the field motion. In other words, the stator field pulls conductors of the rotor, which, however, move with a lower speed than that of the field. The developed torque, T_M, is a product of the rotor radius and sum of electrodynamic forces generated in individual rotor conductors.

When an induction machine operates as a motor, the rotor speed, ω_M, is less than the synchronous velocity, ω_{syn}. The difference of these velocities, given by

$$\omega_{sl} = \omega_{syn} - \omega_M \tag{2.5}$$

and called a *slip velocity*, is positive. Dividing the slip velocity by ω_{syn} yields the so-called *slip*, s, of the motor, defined as

$$s = \frac{\omega_{sl}}{\omega_{syn}} = 1 - \frac{\omega_M}{\omega_{syn}}. \tag{2.6}$$

Here, the slip is positive. However, if the machine is to operate as a generator, in which the developed torque opposes the rotor motion, the slip must be negative, meaning that the rotor must move faster than the field.

2.3 STEADY-STATE EQUIVALENT CIRCUIT

When the rotor is prevented from rotating, the induction motor can be considered to be a three-phase transformer. The iron of the stator and rotor acts as the core, carrying a flux linking the stator and rotor windings, which represent the primary and secondary windings, respectively. The steady-state equivalent circuit of one phase of such a transformer is shown in Figure 2.11. Individual components of the circuit are:

R_s stator resistance
R_{rr} rotor resistance
X_{ls} stator leakage reactance
X_{lrr} rotor leakage reactance
X_m magnetizing reactance
ITR   ideal transformer

The phasor notation based on rms values is used for currents and voltages in the equivalent circuit. Specifically,

$\hat{V}_s$ phasor of stator voltage
$\hat{E}_s$ phasor of stator EMF
$\hat{E}_{rr}$ phasor of rotor EMF
$\hat{I}_s$ phasor of stator current
$\hat{I}_{rr}$ phasor of rotor current
$\hat{I}_m$ phasor of magnetizing current

The frequency of these quantities is the same for the stator and rotor and equal to the supply frequency, f. For formal reasons, it is convenient to

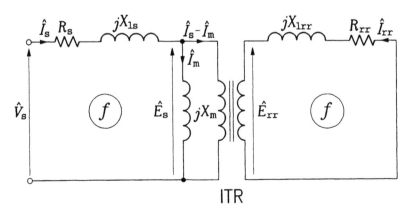

FIGURE 2.11 Steady-state equivalent circuit of one phase of the induction motor at standstill.

assume that both the stator and rotor currents enter the ideal transformer, following a sign convention used in the theory of two-port networks.

When the rotor revolves freely, the rotor angular speed is lower than that of the magnetic flux produced in the stator by the slip speed, ω_{sl}. As a result, the frequency of currents generated in rotor conductors is sf, and the rotor leakage reactance and induced EMF are sX_{lrr} and sE_{rr}, respectively. The difference in stator and rotor frequencies makes the corresponding equivalent circuit, shown in Figure 2.12, inconvenient for analysis. This problem can easily be solved using a simple mathematical trick. Notice that the rms value, I_{rr}, of rotor current is given by

$$I_{rr} = \frac{sE_{rr}}{\sqrt{R_{rr}^2 + (sX_{lrr})^2}}. \tag{2.7}$$

This value will not change when the numerator and denominator of the right-hand side fraction in Eq. (2.7) are divided by s. Then,

$$I_{rr} = \frac{E_{rr}}{\sqrt{\left(\dfrac{R_{rr}}{s}\right)^2 + X_{lrr}^2}}, \tag{2.8}$$

which describes a rotor equivalent circuit shown in Figure 2.13, in which the frequency of rotor current and rotor EMF is f again. In addition, the rotor quantities can be referred to the stator side of the ideal transformer, which allows elimination of this transformer from the equivalent circuit of the motor. The resultant final version of the circuit is shown in Figure

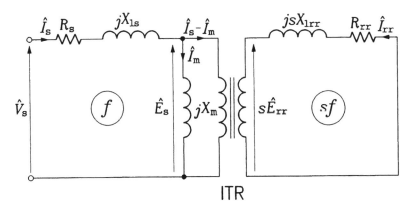

ITR

FIGURE 2.12   Per-phase equivalent circuit of a rotating induction motor with different frequencies of the stator and rotor currents.

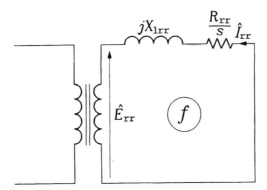

FIGURE 2.13   Transformed rotor part of the per-phase equivalent circuit of a rotating induction motor.

2.14, in which $\hat{E}_r$, $\hat{I}_r$, R_r, and X_{lr}, denote rotor EMF, current, resistance, and leakage reactance, respectively, all referred to stator.

In addition to the voltage and current phasors, time derivatives of magnetic flux phasors are also shown in the equivalent circuit in Figure 2.14. They are obtained by multiplying a given flux phasor by $j\omega$. Generally, three fluxes (strictly speaking, flux linkages) can be distinguished: the stator flux, $\hat{\Lambda}_s$, airgap flux, $\hat{\Lambda}_m$, and rotor flux, $\hat{\Lambda}_r$. They differ from each other only by small leakage fluxes. The airgap flux is reduced in comparison with the stator flux by the amount of flux leaking in the stator; and, with respect to the airgap flux, the rotor flux is reduced by the amount of flux leaking in the rotor.

To take into account losses in the iron of the stator and rotor, an extra resistance can be connected in parallel with the magnetizing reactance.

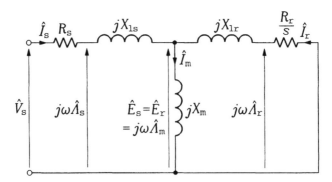

FIGURE 2.14   Per-phase equivalent circuit of the induction motor with rotor quantities referred to the stator.

Except at high values of the supply frequency, these losses have little impact on dynamic performance of the induction motor. Therefore, throughout the book, the iron losses, as well as the mechanical losses (friction and windage), are neglected.

It must be stressed that the stator voltage, $\hat{V}_s$, and current, $\hat{I}_s$, represent the voltage across a phase winding of stator and the current in this winding, respectively. This means that if the stator windings are connected in wye, $\hat{V}_s$ is taken as the line-to-neutral (phase) voltage phasor and $\hat{I}_s$ as the line current phasor. In case of the delta connection, $\hat{V}_s$ is meant as the line-to-line voltage phasor and $\hat{I}_s$ as the phase current.

Although the rotor resistance and leakage reactance referred to stator are theoretical quantities and not real impedances, they can directly be found from simple no-load and blocked-rotor tests. See Section 10.4 for a brief description of these tests.

2.4 DEVELOPED TORQUE

The steady-state per-phase equivalent circuit in Figure 2.14 allows calculation of the stator current and torque developed in the induction motor under steady-state operating conditions. Balanced voltages and currents in individual phases of the stator winding are assumed, so that from the point of view of total power and torque the equivalent circuit represents one-third of the motor. The average developed torque is given by

$$T_M = \frac{P_{out}}{\omega_M}, \tag{2.9}$$

where P_{out} denotes the output (mechanical) power of the motor, which is the difference between the input power, P_{in}, and power losses, P_{loss}, incurred in the resistances of stator and rotor.

The output power can conveniently be determined from the equivalent circuit using the concept of equivalent load resistance, R_L. Because the ohmic (copper) losses in the rotor part of the circuit occur in the rotor resistance, R_r, the R_r/s resistance appearing in this circuit can be split into R_r and

$$R_L = \left(\frac{1}{s} - 1\right)R_r, \tag{2.10}$$

as illustrated in Figure 2.15. Clearly, the power consumed in the rotor after subtracting the ohmic losses constitutes the output power transferred to the load. Thus,

$$P_{\text{out}} = 3R_L I_r^2, \tag{2.11}$$

and

$$T_M = \frac{3R_L I_r^2}{\omega_M}. \tag{2.12}$$

The stator and rotor currents, the latter required for torque calculation using Eq. (2.12), can be determined from the matrix equation

$$\begin{bmatrix} \hat{V}_s \\ 0 \end{bmatrix} = \begin{bmatrix} R_s + jX_s & jX_m \\ jX_m & \dfrac{R_r}{s} + jX_r \end{bmatrix} \begin{bmatrix} \hat{I}_s \\ \hat{I}_r \end{bmatrix}, \tag{2.13}$$

which describes the equivalent circuit in Figure 2.14. Reactances X_s and X_r, appearing in the impedance matrix, are called *stator reactance* and *rotor reactance,* respectively, and given by

$$X_s = X_{ls} + X_m \tag{2.14}$$

and

$$X_r = X_{lr} + X_m. \tag{2.14}$$

An approximate expression for the developed torque can be obtained from the approximate equivalent circuit of the induction motor, shown

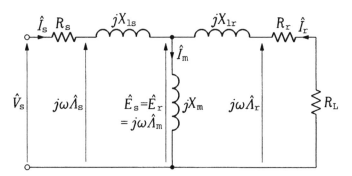

FIGURE 2.15 Per-phase equivalent circuit of the induction motor showing the equivalent load resistance.

in Figure 2.16. Except for very low supply frequencies, the magnetizing reactance is much higher than the stator resistance and leakage reactance. Thus, shifting the magnetizing reactance to the stator terminals of the equivalent circuit does not significantly change distribution of currents in the circuit. Now, the rms value, I_r, of rotor current can be calculated as

$$I_r = \frac{V_s}{\sqrt{\left(R_s + \dfrac{R_r}{s}\right)^2 + X_l^2}},\qquad(2.16)$$

where

$$X_1 = X_{1s} + X_{1r}\qquad(2.17)$$

denotes the total leakage reactance. When I_r, given by Eq. (2.16), is substituted in Eq. (2.12), after some rearrangements based on Eqs. (2.4) and (2.6), the steady-state torque can be expressed as

$$T_M = \frac{1.5}{\pi}\frac{P_P}{f}V_s^2\frac{\dfrac{R_r}{s}}{\left(R_s + \dfrac{R_r}{s}\right)^2 + X_l^2}.\qquad(2.18)$$

The quadratic relation between the stator voltage and developed torque is the only serious weakness of induction motors. Voltage sags in power lines, quite a common occurrence, may cause such reduction in the torque that the motor stalls. The torque-slip relation (2.18) is illustrated in Figure

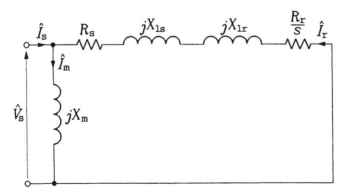

FIGURE 2.16 Approximate per-phase equivalent circuit of the induction motor.

2.17 for various values of the rotor resistance, R_r (in squirrel-cage motors, selection of the rotor resistance occurs in the design stage, while the wound-rotor machines allow adjustment of the effective rotor resistance by connecting external rheostats to the rotor winding). Generally, low values of R_r are typical for high-efficiency motors whose mechanical characteristic, that is the torque-speed relation, in the vicinity of rated speed is "stiff," meaning a weak dependence of the speed on the load torque. On the other hand, motors with a high rotor resistance have a higher zero-speed torque, that is, the starting torque, which can be necessary in certain applications. A formula for the starting torque, $T_{M,st}$, is obtained from Eq. (2.18) by substituting $s = 1$, which yields

$$T_{M,st} = \frac{1.5}{\pi} \frac{P_p}{f} V_s^2 \frac{R_r}{(R_s + R_r)^2 + X_l^2}. \qquad (2.19)$$

The maximum torque, $T_{M,max}$, called a *pull-out torque*, corresponds to a *critical slip*, s_{cr}, which can be determined by differentiating T_M with respect to s and equalling the derivative to zero. That gives

$$s_{cr} = \frac{R_r}{\sqrt{R_s^2 + X_l^2}} \qquad (2.20)$$

and

$$T_{M,max} = \frac{0.75}{\pi} \frac{P_p}{f} \frac{V_s^2}{R_s + \sqrt{R_s^2 + X_l^2}}. \qquad (2.21)$$

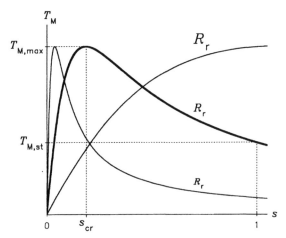

FIGURE 2.17 Torque-slip characteristics of induction motors with various values of the rotor resistance.

It must be reminded that Eqs. (2.16) through (2.21) are based on the approximate equivalent circuit of the induction motor and, as such, they yield only approximate values of the respective quantities.

2.5 STEADY-STATE CHARACTERISTICS

Based on Eqs. (2.3), (2.6), and (2.10) through (2.14), stator current, torque, input and output power, efficiency, and power factor of an induction motor can easily be computed. The input power, P_{in}, efficiency, η, and power factor, PF, can be expressed as

$$P_{in} = 3Re\{\hat{V}_s\hat{I}_s^*\}, \qquad (2.22)$$

$$\eta = \frac{P_{out}}{P_{in}}, \qquad (2.23)$$

and

$$PF = \frac{P_{in}}{S_{in}}, \qquad (2.24)$$

respectively. The apparent input power, S_{in}, in Eq. (2.24) is given by

$$S_{in} = 3V_sI_s, \qquad (2.25)$$

and the P_{in} to S_{in} ratio is equal to the cosine of phase shift between the sinusoidal waveforms of stator voltage and current.

For illustration purposes, a 30-hp induction motor, whose data (some of which have not yet been explained) are listed in Table 2.1, will be used throughout the book. With the rated voltage and frequency, the torque and stator current, input and output power, and efficiency and power factor of this motor are shown in Figures 2.18 through 2.20, respectively. All these variables are plotted as functions of the r/min speed, n. The latter is related to the angular velocity, ω_M, of the motor, expressed in rad/s, as

$$n = \frac{30}{\pi}\omega_M. \qquad (2.26)$$

The rated speed, n_{rat}; torque, $T_{M,rat}$; current, $I_{s,rat}$; and power, P_{rat}, are marked by a broken line to highlight the rated conditions of the motor. The rated torque, current, and powers are much lower than their maximum

TABLE 2.1   Parameters of the Example Motor

Parameter	Symbol	Value
Rated power	P_{rat}	30 hp (22.4 kW)
Rated stator voltage*	$V_{s,rat}$	230 V/ph
Rated stator current**	$I_{s,rat}$	39.5 A/ph
Rated frequency	f_{rat}	60 Hz
Rated slip	s_{rat}	0.027
Rated speed	n_{rat}	1168 r/min
Rated torque	$T_{M,rat}$	183 Nm
Number of pole pairs	p	6
Stator connection		delta
Stator resistance	R_s	0.294 Ω/ph
Stator leakage reactance at 60 Hz	X_{ls}	0.524 Ω/ph
Stator reactance at 60 Hz	X_s	15.981 Ω/ph
Stator inductance	L_s	0.0424 H/ph
Rotor resistance	R_r	0.156 Ω/ph
Rotor leakage reactance at 60 Hz	X_{lr}	0.279 Ω/ph
Rotor reactance at 60 Hz	X_r	15.736 Ω/ph
Rotor inductance	L_r	0.0417 H/ph
Magnetizing reactance at 60 Hz	X_m	15.457 Ω/ph
Magnetizing inductance	L_m	0.041 H/ph
Rotor mass moment of inertia	J_M	0.4 kg.m^2

*The same as the rated voltage, V_{rat}, of the motor, in volts, thanks to the delta connection of the stator. The rated stator voltage, $V_{s,rat}$, in volts per phase, is understood here as the voltage across a phase winding of the stator. In a wye-connected motor, $V_{rat} = \sqrt{3}V_{s,rat}$.

**Not the same as the rated current, I_{rat}, of the motor, in amperes, drawn from the power line. The rated stator current, $I_{s,rat}$, in amperes per phase, is understood here as the current in a single phase of the stator winding. In a wye-connected motor, $I_{rat} = I_{s,rat}$. but in a delta-connected one, $I_{rat} = \sqrt{3}I_{s,rat}$.

values. As seen in Figure 2.20, the rated speed offers the best tradeoff between the efficiency and power factor of the motor.

EXAMPLE 2.1   Calculation of characteristics in Figures 2.18 through 2.20 is elucidated below for the example motor operating with the speed of 1176 r/min at the rated stator voltage of 230 V and frequency of 60 Hz. According to Eq. (2.4), the synchronous speed, n_{syn}, is 120

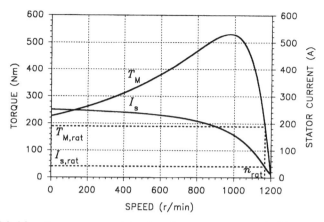

FIGURE 2.18   Stator current and developed torque versus speed of the example motor.

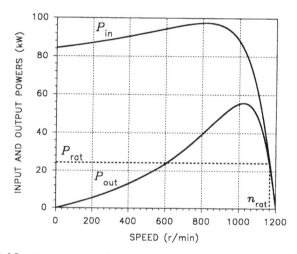

FIGURE 2.19   Input power and output power versus speed of the example motor.

$\times$ 60/6 = 1200 r/min. Thus, from Eqs. (2.6) and (2.10), the slip, s, is $1 - 1176/1200 = 0.02$; and the equivalent load resistance, R_L, is $(1/0.02 - 1) \times 0.156 = 7.644$ Ω/ph. The angular velocity, ω_M, of the motor is $\pi/30 \times 1176 = 123.15$ rad/s.

As in all three-phase systems, the rated stator voltage is given as the *rms value of the line-to-line supply voltage* of the motor. Because stator windings are connected in delta, the same voltage appears across these windings and, consequently, across the input terminals of the per-phase equivalent circuit of the motor. Thus, the rms phasor, $\hat{V}_s$,

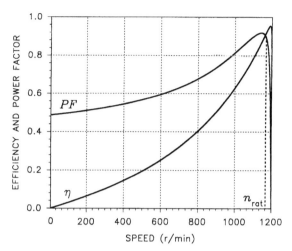

FIGURE 2.20 Efficiency and power factor versus speed of the example motor.

of the stator voltage, taken here as the reference phasor, is 230 V and Eq. (2.13) is

$$\begin{bmatrix} 230 \\ 0 \end{bmatrix} = \begin{bmatrix} 0.294 + j15.981 & j15.457 \\ j15.457 & \dfrac{0.156}{0.02} + j15.736 \end{bmatrix} \begin{bmatrix} \hat{I}_s \\ \hat{I}_r \end{bmatrix}.$$

The popular program MATLAB was used to compute phasors of the stator and rotor currents, yielding $\hat{I}_s$ = 31.15 $\angle$ -0.54 A/ph and $\hat{I}_r$ = 27.41 $\angle$3.06 A/ph. Now, according to Eqs. (2.11) and (2.12), the output power and developed torque can be found as P_{out} = 3 × 7.644 × 27.41^2 = 17229 W and T_M = 17229/123.15 = 139.9 Nm.

The apparent input power to the motor is given by Eq. (2.25) as S_{in} = 3 × 230 × 31.15 = 21494 VA, while the power factor, PF, is equal to the cosine of phase angle of phasor $\hat{I}_s$ of stator current; that is, PF = cos(-0.54) = 0.858. Thus, the real input power, P_{in}, and efficiency, η, are 0.858 × 21494 = 18442 W and 17229/18442 = 0.934, respectively. Check that the same value of P_{in} can be obtained from Eq. (2.22). ■

EXAMPLE 2.2 To evaluate the accuracy of approximate formulas (2.18) through (2.21), exact values of the starting torque, $T_{M,st}$; pullout torque, $T_{M,max}$; rated torque, $T_{M,rat}$; and critical slip, s_{cr}, have been compared with the respective approximate values. The results are

TABLE 2.2   Evaluation of Accuracy of Approximate Formulas (2.18) through (2.21)

Quantity	Exact Value	Eq. (2.18)	Eq. (2.19)	Eq. (2.20)	Eq. (2.21)
Starting torque	227.0 Nm		232.5 Nm		
Pull-out torque	530.9 Nm				549.5 Nm
Rated torque	183.1 Nm	194.5 Nm			
Critical slip	0.187			0.182	

listed in Table 2.2. Good accuracy of the approximate formulas can be observed. The percent errors vary from 2.4% (for the starting torque) to 6.2% (for the rated torque).   ■

It must be pointed out that, in the steady state, induction motors operate only on the negative-slope part of the torque curve, that is, below the critical slip. When the load increases, the resultant imbalance of the motor and load torques causes deceleration of the drive system. This results in an increased motor torque that matches that of the load, ensuring stability of the operation. Conversely, when the load decreases, the motor accelerates until the load torque is matched again.

In Figures 2.18 through 2.20, the motor speed is limited to the 0 to n_{syn} range, n_{syn} denoting the synchronous speed in r/min. This can be translated into the 1 to 0 range of slip. However, in general, an induction machine can operate with any value of slip, positive or negative. In Figure 2.21, the torque and stator current versus speed curves, such as those in Figure 2.18, are extended over the speed range from $-n_{syn}$ to $2n_{syn}$, so that the slip range is 2 to -1. The negative magnitude, I_s, of the stator current at supersynchronous speeds is meant to indicate that the phase shift of the current with respect to the stator voltage is greater than 90° and less than 270°. This implies a negative real power consumed by the motor, that is, the machine operates as a generator. Figure 2.21 illustrates three possible modes of operation of the induction motor: (1) *braking*, with $s > 1$ (i.e., $n < 0$); (2) *motoring*, with $0 < s < 1$ (i.e., $0 < n < n_{syn}$); and (3) *generating*, with $s < 0$ (i.e., $n > n_{syn}$).

In the braking mode, the rotor is forced to rotate against the stator field, which causes high EMFs and currents induced in the rotor conductors. This mode can easily be imposed on a motor by reversing the field, which is accomplished by interchanging two leads between the power line and stator terminals, that is, by changing the phase sequence. However, the braking torque is low, so that this method of slowing the motor down is

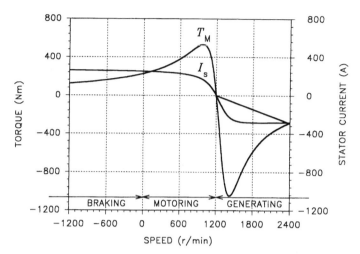

FIGURE 2.21 Torque and current characteristics of the induction motor in a wide speed range.

not very effective. In addition, both the kinetic energy given up by the load and the electric energy supplied to the motor are dissipated in the rotor winding. Thus, no energy is recovered, and the motor is likely to overheat.

Much more efficient braking results from forcing the motor to operate in the generating mode, which requires that the rotor turns faster than the field. This is done by reducing the field speed, n_{syn}, so that it revolves slower than the rotor. According to Eq. (2.4), it can be done by increasing the number, p_p, of pole pairs of the stator or by decreasing the supply frequency, f. Indeed, certain motors have stator windings so arranged that they can be connected in more than one configuration, yielding, for instance, $p_{p,1} = 1$ and $p_{p,2} = 2$. In the adjustable-speed drive systems, the motor is fed from an inverter, which supplies stator currents of variable frequency. There, the generating mode can easily be enforced by keeping track of the rotor speed and reducing the supply frequency accordingly.

2.6 INDUCTION GENERATOR

It has been said that an induction machine rotating with the speed higher than that of the magnetic field of the stator operates as a generator, feeding electrical power back to the supply system. This property is utilized in, for example, induction generators driven by a wind turbine and connected to the grid. The question is whether an induction machine can operate as a stand-alone generator of electric energy. Basically, the answer is negative, because in the absence of magnetic field no EMF can be induced in the rotor.

Yet, stand-alone induction generators are feasible, because to maintain the magnetic field only the reactive power is required. Capacitors connected between the stator terminals of an induction machine can serve as such a source. When the machine is driven by a prime mover, a field buildup is initiated by the residual flux density in stator iron. Analysis of such an induction generator is based on the per-phase equivalent circuit shown in Figure 2.22. The negative resistance R_G represents a source of the input mechanical power, P_{in}. This resistance is a counterpart of the equivalent load resistance, R_L, in Figure 2.15 and given by the same relation, that is,

$$R_G = R_r \left(\frac{1}{s} - 1 \right). \qquad (2.27)$$

For analysis purposes, magnetic saturation of iron of the machine must be taken into account when expressing the stator EMF, E_s, in terms of the magnetizing current, I_m. This can be explained by considering the no-load operation of the generator. No real power is supplied by the rotor, which allows removing the rotor part from the equivalent curcuit in Figure 2.22, yielding the circuit in Figure 2.23. Neglecting the small voltage drop across the stator resistance and leakage reactance, the operating point can be found at the intersection of the magnetization curve, $E_s = f(I_m)$, and the capacitor load line, $V_s = X_C I_C$, where X_C denotes the capacitor impedance, $1/(\omega C)$, and I_C is the capacitor current. As illustrated in Figure 2.24, a too-small capacitance (line 1) is insufficient to provide excitation for the generator, while no solution can be found if the capacitor load

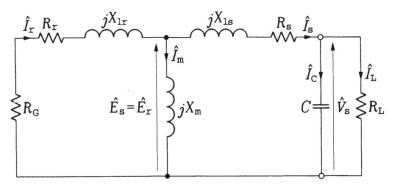

FIGURE 2.22   Per-phase equivalent circuit of the stand-alone induction generator.

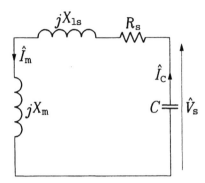

FIGURE 2.23 Per-phase equivalent circuit of the stand-alone induction generator on no load.

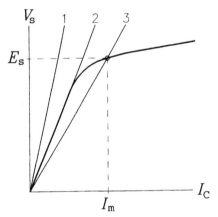

FIGURE 2.24 Determination of the operating point of the induction generator.

line (line 2) coincides with the straight portion of the magnetization curve. On the other hand, a properly selected capacitance (line 3) should not be too large, to avoid unnecessarily high magnetizing and capacitor currents.

It must be stressed that C denotes a *per-phase* capacitance. In practice, the capacitors are connected between each pair of output terminals, that is, they are subjected to the line-to-line voltages. Analyzing the total amount of reactive power associated with these capacitors, it can easily be found that the per-phase capacitance, C, equals the actual capacitance, C_{act}, of the single capacitor only when stator windings are connected in delta (which is the predominant practical connection). With stator windings connected in wye, $C = 3C_{act}$.

The following set of equations can be used for a computer-based analysis of the induction generator operating with the output frequency

ω. A purely resistive load, R_L, is assumed. The load, capacitor, and stator currents, $\hat{I}_L$, $\hat{I}_C$, and $\hat{I}_s$, respectively, are determined as

$$\hat{I}_L = \frac{\hat{V}_s}{R_L}, \tag{2.28}$$

$$\hat{I}_C = -\frac{\hat{V}_s}{jX_C}, \tag{2.29}$$

and

$$\hat{I}_s = \hat{I}_C + \hat{I}_L. \tag{2.30}$$

Now the stator EMF, $\hat{E}_s$, can be found as

$$\hat{E}_s = \hat{V}_s + (R_s + jX_{ls})\hat{I}_s \tag{2.31}$$

and the magnetizing current, $\hat{I}_m$, as

$$\hat{I}_m = f^{-1}(E_s)e^{j(\Theta_E - \frac{\pi}{2})}, \tag{2.32}$$

where Θ_E denotes the angle of phasor $\hat{E}_s$. Finally, the rotor current, $\hat{I}_r$, is given by

$$\hat{I}_r = \hat{I}_m + \hat{I}_s. \tag{2.33}$$

The stator voltage, $\hat{V}_s = V_s$ (reference phasor), in Eqs. (2.28), (2.29), and (2.31) must be such that the balance of reactive powers,

$$X_C I_C^2 = X_{ls}I_s^2 + X_{lr}I_r^2 + E_s I_m^2, \tag{2.34}$$

is satisfied. This, in addition to the nonlinear relation between the stator EMF and magnetizing current, requires an iterative approach to the computations. Once the currents have been found, the balance of real powers,

$$-R_G I_r^2 = R_r I_r^2 + R_s I_s^2 + R_L I_L^2 \tag{2.35}$$

and Eq. (2.27) allow calculation of the slip, s, which is negative, as

$$s = -\frac{R_r I_r^2}{R_s I_s^2 + R_L I_L^2}. \tag{2.36}$$

With the slip known, the rotor angular velocity, ω_M, can be determined as

$$\omega_M = \frac{1 - s}{p_p}\omega \tag{2.37}$$

and the driving torque, T_M, as

$$T_M = \frac{P_{in}}{\omega_M} = -\frac{3R_G I_r^2}{\omega_M},\qquad (2.38)$$

where P_{in} denotes the input power.

EXAMPLE 2.3   It can be shown that when the example motor operates as a stand-alone induction generator with the output frequency of 60 Hz, the per-phase capacitance of 207 μF/ph, and the load resistance of 7.1 Ω/ph, then the stator voltage and output power assume their rated levels of 230 V and 30 hp, respectively. Determine operating conditions of the machine when the load resistance is increased to 10 Ω/ph.

The rated stator voltage, V_s, of 230 V/ph is first assumed, resulting in $\hat{I}_L = 23.0$ A/ph, $\hat{I}_C = 17.9\ \angle 90°$ A/ph, $\hat{I}_s = 29.2\ \angle 38.0°$ A/ph, $\hat{I}_m = 16.0\ \angle -85.6°$ A/ph, and $\hat{I}_r = 24.3\ \angle 4.7°$ A/ph. The left-hand side of Eq. (2.35) turns out to be greater than the right-hand side by 314.7 VA/ph. Gradual increases in V_s reduce this imbalance of reactive powers, until, at $V_s = 249.8$ V/ph, the following solution is reached: $\hat{I}_L = 25.0$ A/ph, $\hat{I}_C = 19.5\ \angle 90°$ A/ph, $\hat{I}_s = 31.7\ \angle 38.0°$ A/ph, $\hat{I}_m = 18.9\ \angle -85.6°$ A/ph, and $\hat{I}_r = 26.4\ \angle 1.5°$ A/ph. The slip, s, of the generator is -0.0167, which corresponds to the rotor speed, n, of 1220 r/min. The input power, P_{in} is 26.7 hp (19.9 kW) and the output power, P_{out}, is 25.1 hp (18.7 kW), yielding the driving torque, T_M, of 146.6 Nm and efficiency of 0.94. Note that the stator voltage is higher than rated, which may be hazardous unless the insulation of stator windings is reinforced.   ∎

Residual magnetism in the iron of a stand-alone induction generator is necessary for the avalanche buildup of the magnetic field when the machine starts turning. The output voltage strongly depends on the load, especially on the reactive component of the load impedance. The slip and, consequently, the output frequency also are load dependent, albeit to a much lesser degree. Usually, the voltage of induction generators is conditioned using power electronic converters.

2.7 SUMMARY

Operation of the induction motor is based on the ingenious principle of induction of EMFs and currents in the rotor that is not directly connected to any supply source. Three-phase currents in stator windings produce a revolving magnetic field, whose angular velocity, called a synchronous velocity of the motor, is proportional to the supply frequency and inversely proportional to the number of pole pairs. The latter parameter, an integer, depends on the configuration of the windings, and it determines the field pattern. The rotor rotates with a speed different than that of the field. Consequently, lines of magnetic flux intersect rotor conductors, inducing the EMFs and currents. Slip, s, which is the relative difference of speeds of the field and rotor, is one of the most important quantities defining operating conditions of an induction machine.

Analysis of the steady-state operation of the induction motor is based on the per-phase equivalent circuit. The mechanical load of the motor is modeled by the equivalent load resistance. The developed torque resulting from interaction between the field and rotor currents strongly depends on the slip. It can be calculated as a ratio of power dissipated in equivalent load resistances of all three phases of the motor to the angular velocity of the rotor. The torque reaches a maximum value, the pull-out torque, at a speed lower than rated. The pull-out torque and the starting torque are higher than the rated torque. Other steady-state characteristics, such as the stator current versus speed, can also be determined from the equivalent circuit.

An induction machine running with a supersynchronous speed operates in the generating mode. Usually, the generating is performed by motors connected to the power system, which provides the reactive power needed for the magnetic field. Stand-alone induction generators are feasible, with capacitors connected across the stator terminals and acting as sources of reactive power.

3

UNCONTROLLED INDUCTION MOTOR DRIVES

In this chapter, operation of uncontrolled induction motor drives is examined. We briefly outline methods of assisted starting, braking, and reversing. Speed control by pole changing is explained, and we describe abnormal operating conditions of induction motors.

3.1 UNCONTROLLED OPERATION OF INDUCTION MOTORS

In a majority of induction motor drives in industrial and domestic applications, the control functions are limited to the turn-on and turn-off and, in certain cases, to assisted starting, braking, and reversing. When driving a load, an induction motor is supplied directly from a power line and operates with fixed values of stator voltage and frequency. The speed of the motor is approximately constant, motors with a stiff mechanical characteristic (i.e., with low dependence of load torque on the speed) having been usually used. As already mentioned, such a characteristic is associated with a low rotor resistance, that is, with low losses in the rotor.

Thus, high-efficiency motors, somewhat more expensive than standard motors, are particularly insensitive to load changes.

Clearly, an uncontrolled motor drive is the cheapest investment, but the lack of speed control carries another price. In many applications, a large percentage of the electric energy is wasted because of that shortcoming. The most common induction motor drives are those associated with fluid transport machinery, such as pumps, fans, blowers, or compressors. To control the flow intensity or pressure of the fluid, valves choking the flow are used. As a result, the motor delivers full power, a significant portion of which is converted into heat in the fluid. This situation is analogous to that of a car driven with a depressed brake pedal. Energy and money savings have been the major reason for the increasing popularity of ASDs, which, typically, are characterized by short payback periods.

Sensitivity to voltage sags constitutes another weakness of uncontrolled drives. Even in highly developed industrial nations such as the United States, the power quality occasionally happens to be poor. Because the torque developed in an induction motor is quadratically dependent on the stator voltage, a voltage sag can cause the motor to stall. This typically leads to intervention of protection relays that trip (disconnect) the motor. Often, the resultant process interruption is quite costly. Controlled drives can be made less sensitive to voltage changes, enhancing the "ridethrough" capability of the motor.

3.2 ASSISTED STARTING

As exemplified in Figure 2.18, the stator current at zero slip, that is, the starting current, is typically much higher than the rated current. Using the approximate equivalent circuit in Figure 2.16, the starting current, $I_{s,st}$, can be estimated as

$$I_{s,st} = \frac{V_s}{\sqrt{(R_s + R_r)^2 + X_I^2}}. \tag{3.1}$$

In the example motor, the starting current, at about 250 A/ph, is 6.3 times higher than the rated current. For small motors this is usually not a serious issue, and they are started by connecting them directly to the power line. However, large motors, especially those driving loads with high inertia or high low-speed torque, require assisted starting. The following are the most common solutions.

1. In *autotransformer starting*, illustrated in Figure 3.1, a three-phase autotransformer is controlled using timed relays. The stator

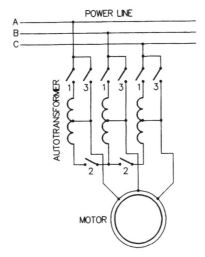

FIGURE 3.1   Autotransformer starting system.

voltage at starting is reduced by shutting contacts 1 and 2, while contacts 3 are open. After a preset amount of time, contacts 1 and 2 are opened and contacts 3 shut.

2. In *impedance starting*, illustrated in Figure 3.2, series impedances (resistive or reactive) are inserted between the power line and the motor to limit the starting current. As the motor gains speed, the impedances are shorted out, first by contacts 1, then by contacts 2.

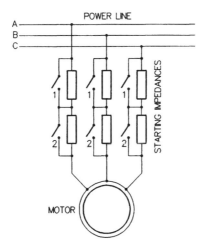

FIGURE 3.2   System with starting impedances.

3. In *wye-delta starting*, illustrated in Figure 3.3, a special switch is used to connect stator phase windings in wye (contacts "w") when the motor is started and, when the motor is up to speed, to reconnect the windings in delta (contacts "d"). With wye-connected phase windings, the per-phase stator voltage and current are reduced by $\sqrt{3}$ in comparison with those for delta-connected windings. The wye-delta switch can be controlled manually or automatically.

4. In *soft-starting*, illustrated in Figure 3.4, a three-phase soft-starter based on semiconductor power switches is employed to reduce the stator current. This is done by passing only a part of the voltage waveform and blocking the remaining part. The volt-

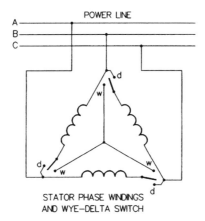

FIGURE 3.3   Starting system with the wye-delta switch.

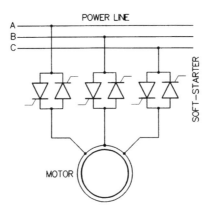

FIGURE 3.4   Soft-starting system.

age and current waveforms are distorted, generating harmonic torques, until, when the motor has gained sufficient speed, the soft-starter connects it directly to the power line. Various starting programs, such as maintaining a constant current or ramping up the voltage, can be realized.

In comparison with the direct online starting, all the preceding methods of assisted starting result in reduction of the starting torque. This, with certain loads, can be a serious disadvantage. As explained later, the variable-frequency starting in ASDs does not have this disadvantage, allowing for high values of the torque.

As an interesting observation, it is worth mentioning that the total energy lost in the rotor during starting is approximately equal to the total kinetic energy of the drive system in the final steady state. This is because the efficiency of power conversion in the rotor is $1 - s$. Again, the variable-frequency starting is superior in this respect, because a low slip is consistently maintained.

3.3 BRAKING AND REVERSING

In drives requiring rapid deceleration, the motor needs to develop a negative torque for braking, especially in systems with low load torque and/or high inertia. Because the torque depends on slip, a proper change in the slip must be effected. Apart from frequency control or changing the number of poles of stator winding, there are two ways to induce a negative torque in an induction machine, *plugging* and *dynamic braking*.

Plugging consists in a reversal of phase sequence of the supply voltage, which is easily accomplished by interchanging any two supply leads of the motor. This results in reverse rotation of the magnetic field in the motor; the slip becomes greater than unity and the developed torque tries to force the motor to rotate in the opposite direction. If only stopping of the drive is required, the motor should be disconnected from the power line at about the instant of zero speed.

Plugging is quite a harsh operation, because both the kinetic energy of the drive and input electric energy must be dissipated in the motor, mostly in the rotor. This braking method can be compared to shifting a transmission into reverse to slow down a running car. The total heat produced in the rotor is approximately three times the initial energy of the drive system. Therefore, plugging must be employed with caution to avoid thermal damage to the rotor. Low-inertia drives and motors with high rotor resistance and, therefore, with a large high-slip torque (see Figure 2.17) are the best candidates for effective plugging.

EXAMPLE 3.1   To illustrate braking by plugging, consider the example motor driving a load under rated operating conditions. The mass moment of inertia of the load is twice that of the motor. The initial braking torque and total energy dissipated in the rotor by the time the motor stops are to be determined.

The rated speed is 1168 r/min. Thus, when the speed of magnetic field is reversed, the initial slip, s, is $(1200 + 1168)/1200 = 1.973$. The matrix equation (2.13) is

$$\begin{bmatrix} 230 \\ 0 \end{bmatrix} = \begin{bmatrix} 0.294 + j15.981 & j15.457 \\ j15.457 & \dfrac{0.156}{1.973} + j15.736 \end{bmatrix} \begin{bmatrix} \hat{I}_s \\ \hat{I}_r \end{bmatrix},$$

and, when solved, it yields $I_s = 261.3$ A/ph and $I_r = 256.7$ A/ph. The rotor velocity, ω_M, is $\pi \times 1168/30 = 122.3$ rad/s and the equivalent load resistance, R_L, found from Eq. (2.10), is -0.077 Ω/ph. It is negative, because $s > 1$, and consequently the developed torque T_M, as calculated from Eq. (2.12), is negative too. Specifically,

$$T_M = \frac{3 \times (-0.077) \times 256.7^2}{122.3} = -124.5 \ Nm,$$

which is only two-thirds of the rated torque, while the stator current is 6.6 times the rated value. The maximum braking torque using this method occurs at zero speed and equals the starting torque of 227 Nm (see Table 2.2). The corresponding stator current of 250 A/ph (see Section 3.2) is still very high at 6.3 times the rated current.

The load mass moment inertia is $2 \times 0.4 = 0.8$ kg.m^2, and the energy, E_r, dissipated in the rotor is three times the initial kinetic energy of the drive system. Thus,

$$E_r = 3\frac{(J_M + J_L)\omega_M^2}{2} = 3\frac{(0.4 + 0.8)122.3^2}{2} = 26,923 \ J. \quad \blacksquare$$

Dynamic braking is realized by circulating direct current in stator windings. For braking, the motor is disconnected from the power line, and any two of its phases are connected to a dc voltage source. The dc stator current produces a stationary magnetic field, so that ac EMFs and currents are induced in the rotor bars, and a braking torque is developed. The braking torque, $T_{M,br}$, is given by the approximate equation

$$T_{M,br} = 3\left(\frac{X_m I_{s,dc}}{\omega_{syn}}\right)^2 \frac{R_r \omega_M}{R_r^2 + \left(\dfrac{\omega_M}{\omega_{syn}} X_m\right)^2}, \tag{3.2}$$

where $I_{s,dc}$ denotes the dc stator current. The relation between the braking torque and motor speed, n_M, resembles that for supersynchronous speeds (see Figure 2.22), with the maximum braking torque in the vicinity of $n_M = n_{syn} R_r / X_m$. Indeed, with the stationary field, a braking motor can be thought of as running at a supersynchronous speed. Although no energy regeneration is possible, the amount of heat dissipated in the rotor is one-third of that for plugging, being approximately equal to the initial kinetic energy of the drive system.

The dynamic-braking arrangement is illustrated in Figure 3.5. The braking dc current encounters only the stator resistance, so the dc source supplying this current must have voltage much lower than the rated ac voltage of the motor. Therefore, a step-down transformer is used, the reduced secondary ac voltage of which is converted into dc voltage by a diode rectifier. Normally, the motor operates with contacts 1 closed and contacts 2 and 3 opened. For braking, the motor is disconnected from the power line by opening contacts 1, and two of its phases are connected to the rectifier by closing contacts 2. Contacts 3 are closed simultaneously, providing power supply for the transformer. In large motors, instead of

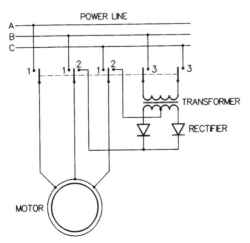

FIGURE 3.5   System for the dynamic braking.

the single-phase transformer and rectifier in Figure 3.5, their three-phase counterparts can be used.

EXAMPLE 3.2   To compare dynamic braking with braking by plugging, the motor from Example 3.1 is analyzed when disconnected from the ac line and connected to a dc source. The dc stator current is twice the rms-rated current of the motor.

The dc stator current, $I_{s,dc}$, is $2 \times 39.5 = 79$ A, which allows us to determine the required voltage, $V_{s,dc}$, of the dc source as $V_{s,dc} = 2R_s I_{s,dc} = 2 \times 0.294 \times 79 = 46.5$ V. This is about one-fifth the rated ac stator voltage, which confirms the need for the step-down transformer in the system in Figure 3.5.

The synchronous angular velocity, ω_{syn}, of the motor is $\pi \times 1200/30 = 125.7$ rad/s, and the braking torque, $T_{M,br}$, at the initial velocity, ω_M, of 122.3 rad/s (see Example 3.1) is calculated from Eq. (3.2) as

$$T_{M,br} = 3\left(\frac{15.457 \times 79}{125.7}\right)^2 \frac{0.156 \times 122.3}{0.156^2 + \left(\dfrac{122.3}{125.7}15.457\right)^2} = 23.9 \; Nm.$$

This is a very low value, only 13% of the rated torque, but the braking torque increases rapidly with the decreasing speed of the motor. Because $R_r/X_m \approx 0.01$, the maximum braking torque, $T_{M,br(max)}$, occurs at the motor velocity of $0.01\omega_{syn}$, that is, at $\omega_M = 1.257$ rad/s. Then, using Eq. (3.2) again,

$$T_{M,br(max)} = 3\left(\frac{15.457 \times 79}{125.7}\right)^2 \frac{0.156 \times 1.257}{0.156^2 + \left(\dfrac{1.257}{125.7}15.457\right)^2}$$

$$= 1{,}151 \; Nm,$$

which is 6.3 times the rated torque and more than twice the pull-out torque (see Table 2.2). Generally, the lower the R_r/X_m ratio, the higher the ratio of the maximum braking torque to that at the rated speed.

The energy, E_r, dissipated in the rotor equals the initial kinetic energy of the drive system, that is, it is only one-third of that when plugging is used. Based on results of Example 3.1, $E_r = 26923/3 = 8974$ J. The comparison of plugging and dynamic braking has shown definite superiority of the latter method. The average braking torque is much higher than with plugging, and the heat generated in the motor, both in stator and rotor, is much lower. ∎

Certain drives require prolonged stopping. For instance, too-rapid speed reduction of a conveyor belt could cause spillage, and that of a centrifugal pump may result in pipe damage due to the water-hammer effect. In such cases, power electronic soft-starters can be used to slowly reduce (ramp down) the stator voltage.

Reversing an induction motor drive involves braking the motor and restarting it in the opposite direction. The braking and starting can be done in any of the ways described above. Plugging is a good option for motors running light, while simply disconnecting the motor from the power line can be sufficient for quick stopping of drives with a high reactive load torque. In some drives, the reversing is performed in the gear train so that the motor operation is not affected.

3.4 POLE CHANGING

A formula for speed, n_M, of the induction motor as a function of the supply frequency, f, number of pole pairs, p_p, of the magnetic field, and slip, s, of the motor can be obtained from Eqs. (2.4) and (2.6) as

$$n_M = 60\frac{f}{p_p}(1 - s). \tag{3.3}$$

On the other hand, with a fixed output power, the speed is inversely proportional to the developed torque [see Eq. (2.9)]. Therefore, observing two motors of the same power, frequency, and voltage ratings, of which one has a two-pole stator winding and the other a four-pole winding, and which drive identical loads, the four-pole machine would rotate with half of the speed of the two-pole one but with twice as high a torque. Thus, a motor with p_p pole pairs is equivalent to a two-pole machine connected to the load through gearing whose gear ratio, N, as defined by Eq. (1.4), is $1/p_p$. The gear-ratio property of the number of poles is utilized in certain motors for speed control. Such motors have stator windings so constructed that they can be connected in various arrangements, in order to produce magnetic fields of an adjustable pole number, for instance two, four, and six. In this way, the synchronous speed can assume several distinct values, such as 3600 r/min, 1800 r/min, and 1200 r/min.

The topic of stator windings in ac machines is vast, and it exceeds the scope of this book. Interested readers are referred to relevant sources, for instance the excellent manual by Rosenberg and Hand, 1986, which can be found in the Literature section at the end of this book. Here, only one example of pole changing is illustrated in Figure 3.6. It shows a four-coil winding of phase A, which can be connected to produce a four- or eight-pole magnetic field. In the four-pole arrangement seen in Figure 3.6(a), terminals x and y are shorted forming one end of the winding, while terminal z makes up the other end. When, as in Figure 3.6(b), x and y are disconnected from each other and used as ends of the winding, an eight-pole field is generated.

Arrangement of stator windings affects the developed torque, because the torque is dependent on stator current, which, in turn, depends on the stator impedance. These dependencies allow better matching of a motor to the load. For instance, when a two-pole stator is reconnected to four-pole operation, the resulting pull-out torque can be the same as before (constant torque connection), half of its previous value (square-law torque connection), or twice its previous value (constant power connection). Clearly, these three types of torque-speed relationship are most suitable for loads with the constant, positive, and negative coefficient k in Eq. (1.12), respectively (see Figure 1.1).

3.5 ABNORMAL OPERATING CONDITIONS

Abnormal operation of an induction motor drive may be caused by internal or external problems. The most common electrical and mechanical faults in the motor are:

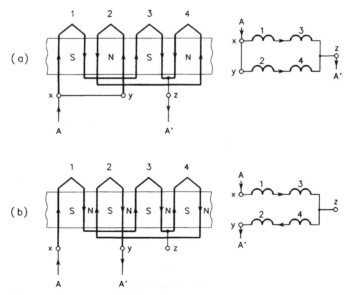

FIGURE 3.6   Pole changing: (a) four-pole stator winding, (b) eight-pole stator winding.

1. Short circuit in the stator winding, which can occur between turns of the same phase (interturn fault), between different phases (interphase fault), or between a phase winding and ground (ground fault). Serious stator faults cause the overcurrent protection circuits to react immediately, but minor faults take time to spread.
2. Cracked rotor bars, resulting from frequent thermal and mechanical stresses, for instance in often-started motors. The cracking usually occurs at the junction with the end ring. The damage to the rotor reduces the torque of the motor and introduces low-frequency harmonic torques. Healthy bars must then carry an increased load, so they are likely eventually to crack too.
3. Bearing failures, caused by wear and accelerated by such mechanical imperfections as rotor unbalance and eccentricity or misalignment of the motor and load shafts.

Interestingly, the incidence of stator and bearing faults in induction motors in adjustable-speed drive systems has been found to be significantly higher than that in uncontrolled drives. It turns out that the switching operation of power electronic inverters supplying the motors in variable-frequency drives causes increased voltage stresses on stator insulation and, in certain cases, microsparking in the bearings.

External factors that may cause abnormal operation of the induction motor are:

1. Poor voltage quality, such as sags or unbalance. As already explained in Section 2.4, voltage sags result in quadratic reduction of the developed torque, so that the motor may stall. Voltage unbalance produces harmonic torques, and it increases losses in the motor. High losses and poor power factor also occur when the stator voltage is too high. Generally, the voltage is considered to be of good quality when it does not strain from the rated value by more than $\pm 10\%$, from the rated frequency by more than $\pm 5\%$, and from ideal balance by more than $\pm 2\%$.
2. Phase loss, typically resulting from the action of protection relays in the power system or fuses in the supply line. Basically, the induction motor can run on two phases, albeit with a significantly reduced torque and increased stator current. Eventually, thermal overload relays will trip the circuit breaker and disconnect the motor from the line.
3. Mechanical overload, which may cause overheating or even stalling the motor. Although induction motors have a significant torque margin (see Figure 2.18), prolonged operation with

overload is hazardous and prevented by the already mentioned thermal overload relays.

3.6 SUMMARY

Most induction motors in industrial and household use operate in an uncontrolled manner, being supplied directly from the power system. The motor speed is roughly constant, but the developed torque is sensitive to changes in the stator voltage. In many applications, significant energy savings could be realized by replacing the uncontrolled drive with an adjustable-speed one.

Motors with difficult starting conditions require means for assisted starting so that the motor does not overheat. Autotransformers, series impedances, wye-delta switches, or electronic soft-starters are commonly employed.

For assisted braking, when the load torque alone is insufficient to quickly stop the motor, plugging or dynamic braking can be applied. Plugging, consisting in the reversal of magnetic field in the motor by changing the phase sequence, is a harsh operation because the accompanying energy losses in the rotor amount to three times the initial kinetic energy of the drive. Moreover, the braking torque in most motors is relatively low. Dynamic braking, utilizing a dc stator current to produce a stationary field, offers better operating conditions, but it requires a step-down transformer and a rectifier. Prolonged stopping can be realized using soft-starters. Reversing an uncontrolled drive involves stopping and restarting. Plugging can be used for motors running light. The gear train itself can be of a reversible type.

Pole changing allows two or more different synchronous speeds in specially constructed motors. The stator windings are so arranged that they can be switched into configurations producing various patterns of the magnetic field.

Abnormal operating conditions of induction motor drives can be of internal or external origin. Motor faults can be electrical or mechanical. Most common faults are short circuits in the stator, cracked rotor bars, and bearing failures. Poor quality of the supply voltage, phase loss, and mechanical overloads are the most common external causes of aggravated operation of induction motors.

4

POWER ELECTRONIC CONVERTERS FOR INDUCTION MOTOR DRIVES

In this chapter, we review power electronic converters used in ASDs with induction motors. Various types of rectifiers providing the dc supply voltage for inverters feeding the motors are presented, and we describe voltage source inverters, including three-level and soft-switching inverters, and current source inverters. Control methods for inverters, with a stress on the use of voltage space vectors, are illustrated. Finally, we outline undesirable side effects of the switching operation of power converters.

4.1 CONTROL OF STATOR VOLTAGE

As seen from Eq. (3.3), the speed of an induction motor can be controlled by changing the number of poles, slip, and the supply frequency. The pole changing has already been described, and, if the motor has that capability, it only requires an appropriate switch. Changes of slip can be effected by varying the stator voltage, particularly in motors with soft mechanical characteristics. However, this method is inefficient, because rotor losses are proportional to the slip. Also, in most motors, it is ineffective because of the narrow range of controllable slip (from zero to the critical value). For wide-range speed control, adjusting the supply frequency constitutes the only practical solution. The frequency control must be accompanied by magnitude control of the stator voltage.

To produce adjustable-frequency, adjustable-magnitude, three-phase voltage for induction motor drives, power electronic *inverters* are most commonly used. Inverters are dc to ac converters, so the regular 60-Hz (50-Hz in many countries) ac voltage must first be *rectified* to provide the dc supply for the inverter. Much less common are *cycloconverters*, which operate directly on the 60-Hz supply, but whose output frequency is inherently much lower than the input (supply) frequency. They are mostly employed in high-power synchronous motor drives. The soft-starters described in Section 3.2 are based on *ac voltage controllers*, which are ac to ac converters with adjustable rms value of the output voltage. The frequency is not changed in ac voltage controllers; that is, the output voltage has the same frequency as the supply voltage. Because of their marginal use in induction motor drives, cycloconverters and ac voltage controllers are not covered in this book.

In the subsequent sections, rectifiers and inverters employed in ASDs with induction motors are briefly described. Operating principles of inverters, knowledge of which is needed for in-depth understanding of control methods for induction motors, are particularly stressed. This necessarily sketchy information may be insufficient for readers with no background in power electronics, for whom studies of relevant literature are strongly recommended. See, for example, *Introduction to Modern Power Electronics*, 1998, by this author.

4.2 RECTIFIERS

Rectifiers in induction motor ASDs supply dc voltage to inverters. The three-phase full-wave (six-pulse) diode rectifier, shown in Figure 4.1, is most commonly employed. At any time, only two out of six diodes conduct the output current, i_o. These are the diodes, subjected to the highest line-to-line input voltage. For instance, if at a given instant the highest line-to-line voltage is v_{AB}, diodes DA and DB' conduct the output current, so that $i_A = i_o$ and $i_B = -i_o$. The other four diodes are then reverse biased, while the output voltage, v_o, equals v_{AB}.

Because, thanks to the conducting diodes, the highest line-to-line input voltage appears at the output of the rectifier, the output voltage is the envelope of all six line-to-line voltages of the supply line. This is

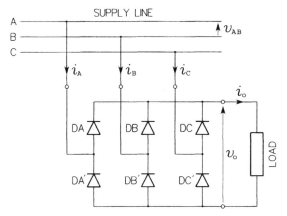

FIGURE 4.1   Six-pulse diode rectifier.

illustrated in Figure 4.2, which shows the line-to-line voltages and output voltage of the six-pulse diode rectifier. The output voltage is not ideally of the dc quality, but it has a high dc component, V_o (average value of v_o), given by

$$V_o = \frac{3}{\pi}V_{LL,m} \approx 0.955V_{LL,m}, \qquad (4.1)$$

where $V_{LL,m}$ denotes the peak value of line-to-line input voltage. The output current, whose example waveform is also shown in Figure 4.2, depends on the load, but its waveform has even less ripple (ac component) than does the voltage waveform.

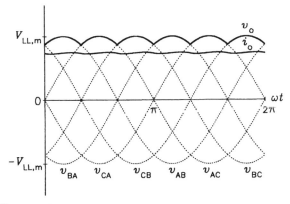

FIGURE 4.2   Waveforms of the output voltage and current in a six-pulse diode rectifier.

Assuming an ideal dc output current, $i_o = I_o$, the input line current have a rectangular waveform, as shown in Figure 4.3. This is a serious annoyance for the supplying power system, which is designed to operate with sinusoidal voltages and currents. The high harmonic content of the square-wave current drawn by the rectifier and the resultant distortion of the voltage waveforms in the power system cause interference with operation of sensitive communication equipment, and they may precipitate unwarranted intervention of system protection circuits. It should be pointed out that, thanks to the half-wave symmetry of the current waveform, no even (2nd, 4th, etc.) harmonics are present, while the three-phase balance of currents in individual supply wires causes the absence of triple harmonics (3rd, 9th, etc.) as well. As a result, the low-order harmonics in the input current of a diode rectifier are the 5th, 7th, 11th, 13th, etc.

When the diodes are replaced with SCRs (silicon controlled rectifiers, also known as thyristors), TA through TC′, as in Figure 4.4, a phase-controlled six-pulse rectifier is obtained. The adjustable dc output voltage, V_o, of the rectifier is given by

$$V_o = \frac{3}{\pi}V_{LL,m}\cos(\alpha_f), \qquad (4.2)$$

where α_f denotes the firing angle, which determines instants of turning on (firing) the SCRs. Specifically, $\alpha_f = 0$ represents the situation when an SCR is fired at the same instant at which the respective diode in an uncontrolled rectifier in Figure 4.1 would begin conducting. If, in the ωt domain, the firing is delayed with respect to that instant by α_f radians, the average output voltage is reduced in proportion to $\cos(\alpha_f)$. Waveforms

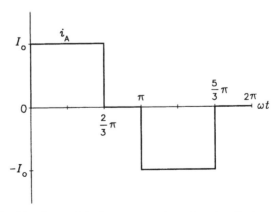

FIGURE 4.3 Waveform of the input current in a six-pulse diode rectifier.

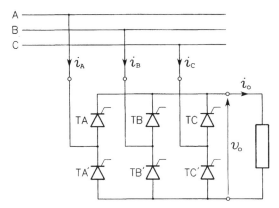

FIGURE 4.4   Six-pulse phase-controlled rectifier.

of the output voltage, v_o, and current, i_o, with the firing angle of 45° are illustrated in Figure 4.5.

Eq. (4.2) implies a possibility of negative dc output voltage when the firing angle exceeds 90°. Because the output current cannot be negative (it would have to flow from the cathode to anode of the SCRs), the negative value of V_o indicates transfer of power from the load to the supply system. Clearly, that requires an active load, such as an electric machine, capable of delivering electrical energy. A controlled rectifier transferring power from the load to the supply is said to operate in the *inverter mode*. This mode is illustrated in Figure 4.6, which shows the output voltage and current waveforms for the firing angle of 135°.

Input current waveforms in the controlled rectifier are similar to those in the diode rectifier. In addition to the rectangular waveform of these

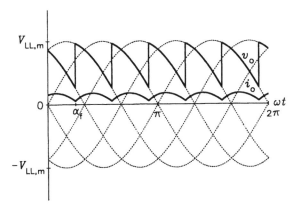

FIGURE 4.5   Waveforms of the output voltage and current in a six-pulse phase-controlled rectifier ($\alpha_f = 45°$).

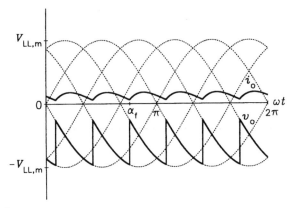

FIGURE 4.6 Waveforms of the output voltage and current in a six-pulse phase-controlled rectifier in the inverter mode ($\alpha_f = 135°$).

currents, shown in Figure 4.7, the input power factor of the controlled rectifier is lower than that of the diode rectifier. The power factor, which similarly to the dc output voltage is proportional to the cosine of the firing angle, decreases with the increase of this angle. The poor quality of currents drawn from the power system is a major disadvantage of uncontrolled and phase-controlled rectifiers.

The problem of harmonic pollution of the power system caused by power electronic converters, often called *nonlinear loads*, is very serious, and significant efforts to combat the system harmonics are being made. The most common solution is to install appropriate filters, either between the power system and the offending converter (series filters) or in parallel with the converter (parallel filters). Filters can be passive or active. Passive

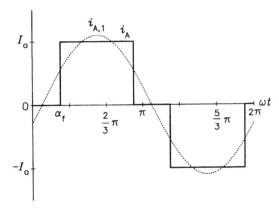

FIGURE 4.7 Waveform of the input current in a six-pulse phase-controlled rectifier.

filters are simple LC (inductive-capacitive) circuits designed to block and shunt current harmonics so that they are drawn from filter capacitors rather than from the power system. With respect to diode rectifiers, the so-called *harmonic traps* are often used. They are series-resonant LC circuits, tuned to frequencies of the lowest harmonics of the input current, for instance the 5th, 7th, 11th, and 13th. The harmonic traps shunt the respective harmonic currents from the power system. The remaining, unfiltered harmonics usually have such low amplitudes that waveforms of currents drawn from the system are close to ideal sinusoids.

The resonant frequencies of harmonic traps are relatively low, because even the 13th harmonic has a frequency well below 1 kHz. Therefore, the inductors and capacitors used in the traps are large and expensive. To significantly reduce the size of passive filters, pulse width modulated (PWM) rectifiers must be used. There are two types of these converters, the voltage source and current source PWM rectifiers.

The *voltage source PWM rectifier,* based on IGBTs, the most popular semiconductor power switch nowadays (the so-called *non-punch-through* IGBTs must be used because of the ac input voltages), is shown in Figure 4.8. The three-phase line with input filters based on inductors L_i and capacitors C_i constitutes the voltage source for the rectifier. The input inductors do not have to be physical components, because the supplying power system itself may possess sufficient inductance, but the capacitors are necessary. The output inductance, L_o, which can be provided by the

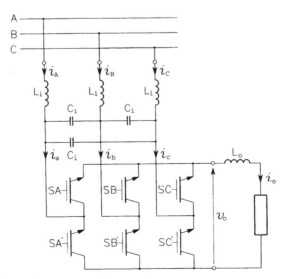

FIGURE 4.8   Voltage source PWM rectifier.

load, smooths the output current. Switches, SA through SC′, of the rectifier are turned on and off many times per cycle of the input voltage in such a way that the fundamental input currents follow desired reference values. Example waveforms of the output voltage, v_o, and current, i_o, of the rectifier are shown in Figure 4.9, and those of the input current, i_a, and its fundamental, $i_{a,1}$, in Figure 4.10. The fundamentals are supplied from the power line, while the high-frequency harmonic components of the pulsed currents, i_a, i_b, and i_c, are mostly drawn from the capacitors. As a result, waveforms of currents i_A, i_B, and i_C, supplied by the power

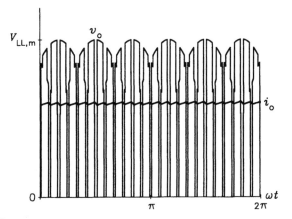

FIGURE 4.9 Waveforms of the output voltage and current in a voltage source PWM rectifier.

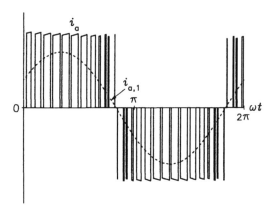

FIGURE 4.10 Waveforms of the input current and its fundamental in a voltage source PWM rectifier.

system and shown in Figure 4.11, are close to ideal sinusoids, with only a small amount of ripple.

The dc output voltage of the voltage source rectifier cannot be adjusted to a value greater than the peak value of line-to-line supply voltage. In contrast, the *current source PWM rectifier* shown in Figure 4.12 allows the boosting of the output voltage. The current source properties of the rectifier result from the input inductors, L_i. Because the rectifier switches provide direct connection between the input and output of the converter, the output capacitor, C_o, is necessary to prevent connecting the input inductance, carrying certain current, with the load inductance, which may conduct a different current. The same capacitor smooths the output voltage,

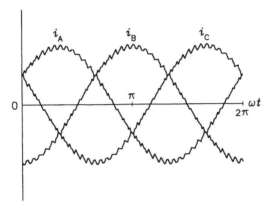

FIGURE 4.11 Waveforms of currents supplied by the power system to the voltage source PWM rectifier.

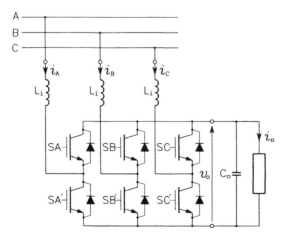

FIGURE 4.12 Current source PWM rectifier.

v_o. Analogously, should a smoothing capacitor be used in the voltage source PWM rectifier in Figure 4.8, a physical inductor L_i would have to be used between the rectifier output and the capacitor to avoid connecting this capacitor, charged to a certain voltage, across the input capacitor charged to a different voltage.

The semiconductor power switches are paired with inverse-parallel freewheeling diodes, which provide alternative paths for currents that cannot flow through switches. Suppose, for example, that switch SA' is turned on and conducts current i_A, whose polarity is that shown in Figure 4.12. When the switch is turned off, the current cannot change instantly, having been maintained by the input inductor in phase A. As a result, the current will force its way through the freewheeling diode of switch SA. Thanks to the output capacitor, the output voltage and current waveforms are practically of the dc quality, with a minimal ripple. Currents drawn from the power system are similar to those in the voltage source PWM rectifier (see Figure 4.11).

The phase-controlled and PWM rectifiers have the capability of reversed power flow, necessary for efficient operation of the drive system in the second and fourth quadrants. In practice, multiquadrant drives are much less common than the single-quadrant ones, which explains the already-mentioned dominance of diode rectifiers in induction motor ASDs. PWM rectifiers are mostly used in low- and medium-power drive systems, with phase-controlled rectifiers employed in the higher ranges of power.

4.3 INVERTERS

The three-phase *voltage source inverter* (VSI) is shown in Figure 4.13. The voltage source for the inverter is made up from a rectifier and the so-called dc link, composed of a capacitor, C, and inductor, L. If the ac machine fed from the inverter operates as a motor (i.e., in the first or third quadrant), the *average* input current is positive. However, the *instantaneous* input current, i_i, may assume negative values, absorbed by the dc-link capacitor which, therefore, is necessary. The capacitor also serves as a source of the high-frequency ac component of i_i, so that it is not drawn from the power system via the rectifier. In addition, the dc link capacitor smooths and stabilizes the voltage produced by the rectifier. The optional dc-link inductor is less important, being introduced to provide an extra screen for the power system from the high-frequency current drawn by the inverter.

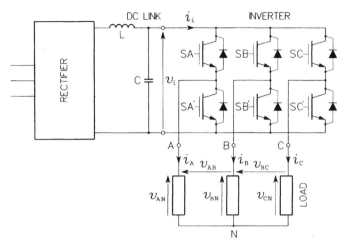

FIGURE 4.13   Voltage source inverter.

Clearly, the topology of the voltage source inverter represents an inverse of that of the current source PWM rectifier in Figure 4.12 (note that the load of the inverter contributes the inductances corresponding to input inductances, L_i, of the rectifier). Here, the freewheeling diodes provide alternative paths for output currents. Both semiconductor power switches in a given leg (phase) of the inverter may not be on simultaneously, because they would short the input terminals. On the other hand, with both switches off, the output voltage would be indeterminable, because the potential of the respective output terminal would depend on which diode is conducting the output current in that phase. This would make the open-loop control of the output voltage impossible. Therefore, voltage source inverters are so controlled that one switch in each leg is on and the other is off. In this way, the turned-on switch connects one of the input terminals to the output terminal, and potentials of all three output terminals are always known. To avoid the so-called *shot-through*, that is, potentially damaging simultaneous conduction of both switches in the same leg, turn-on of a switch is delayed a little with respect to turn-off of the other switch. This delay, on the order of few microseconds, is called a *dead time* or *blanking time*.

The voltage source inverter can operate in both the PWM mode and the so-called *square-wave mode*, characterized by rectangular waveforms of the output voltage. The square-wave operation yields the highest voltage gain of the inverter, but the quality of output current is poorer than that in the PWM mode.

Figure 4.14 shows the *current source inverter* (CSI) which, in the square-wave mode, produces rectangular waveforms of the output current. For consistency, IGBTs are shown here as the inverter switches, but practical current source inverters are often of such a high power that they must be based on GTOs or SCRs with commutating circuits (to turn the SCRs off). The current-source supply is provided by a controlled rectifier with closed-loop current control and the inductive dc link. The inverter differs from its voltage source counterpart by the absence of freewheeling diodes, which are unnecessary because the constant input current is never negative.

Addition of capacitors at the output allows for PWM operation of the current source inverter. These capacitors are marked in Figure 4.14 using broken lines. The switching action of inverter switches results in pulsed waveforms of currents i_a, i_b, and i_c, but the capacitors shunt most of the high-frequency harmonic content of these currents. Thus, waveforms of the output currents, i_A, i_B, and i_C, are rippled sinusoids. This inverter is an inverse of the voltage-source rectifier in Figure 4.8.

Recently, *multilevel voltage source inverters* have been receiving serious attention. They allow for higher voltage ratings than the classic inverter in Fig. 4.12 and produce currents of higher quality, albeit at the expense of a higher device count. The number of levels is meant as the number of values of the voltage at an output terminal of the inverter. For instance, in the voltage source inverter described before, each terminal can be

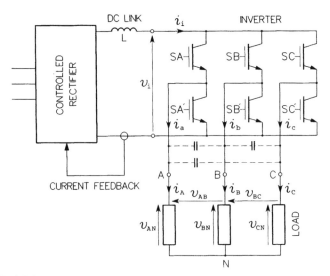

FIGURE 4.14 Current source inverter.

connected to either the positive or negative dc bus and the inverter can therefore be termed as a two-level inverter. The most common, so-called *neutral-clamped*, three-level inverter is shown in Figure 4.15. Each leg of the inverter is composed of four semiconductor power switches, S_1 through S_4, with freewheeling diodes, D_1 through D_4, and two clamping diodes, D_5 and D_6, that prevent the dc-link capacitors from shorting. The dc link is based on two input capacitors, C_i, forming a capacitive voltage divider and an input inductor, L_i (optional).

Although the presence of 12 semiconductor power switches in the three-level inverter implies a very high number of possible inverter states, only 27 states are employed in practice. Specifically, each leg of the inverter is allowed to assume one of the three following states: (1) S1 and S2 are on, S3 and S4 are off; (2) S2 and S3 are on, S1 and S4 are off; and (3) S1 and S2 are off, S3 and S4 are on. It can be seen that the dc input voltage, V_i, is always applied to a pair of series-connected switches. Therefore, the voltage rating of a three-level inverter can be twice as high as the rated voltage of the switches.

All the inverters described are characterized by *hard switching*, that is, each semiconductor switch turns off while carrying a nonzero current and turns on under a nonzero voltage. As a result, each switching is associated with certain energy loss. High switching frequencies, necessary for high quality of the output current, reduce efficiency of the inverter. Side

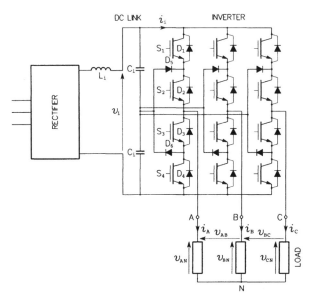

FIGURE 4.15   Three-level neutral-clamped inverter.

effects of hard switching, such as the radiated electromagnetic interference (EMI) or overvoltages in long cables connecting the inverter to the load, are also a problem. Therefore, a great amount of research and development effort has been devoted to *soft-switching* inverters. Generally, two types of soft switching can be distinguished, zero-voltage switching (ZVS) and zero-current switching (ZCS). Many topologies of soft-switching inverters have been developed within the last decade.

The classic *resonant dc link* (RDCL) *inverter* shown in Figure 4.16 employs the ZVS principle. The supply rectifier with the dc link capacitor C_1 and inductor L_1 (optional) constitute the dc voltage source for the inverter. Inductor L_2 and capacitor C_2 form a resonant circuit, the resonance in which is triggered by simultaneous turn-on of both switches in a leg of the inverter, followed by turn-off. When voltage across capacitor C_2 reaches zero, ZVS conditions are created for inverter switches. The active clamp based on the capacitor C_3 and switch S is used to clip the resonant pulses of the output voltage. Without the clamp, the voltage pulses, whose amplitude largely exceeds the dc supply voltage, would impose unreasonably high voltage rating requirements on the switches and diodes of the inverter.

One phase (phase A) of the *auxiliary resonant commutated pole* (ARCP) *inverter* is shown in Figure 4.17. Capacitors C_1 and C_2 form the dc link. To minimize high dynamic stresses on main switches, SA and SA$'$, a resonant snubber, based on the inductor L_A and capacitors C_{A1} and C_{A2}, is employed. The resonance is triggered by turning on the

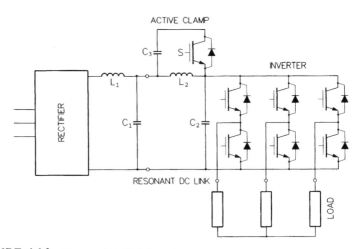

FIGURE 4.16 Resonant dc link inverter.

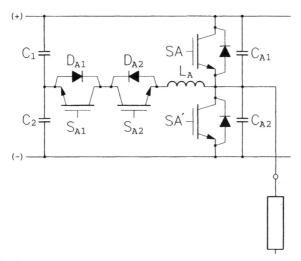

FIGURE 4.17 One phase of the auxiliary resonant commutated pole inverter.

bidirectional switch, composed of auxiliary switches S_{A1} and S_{A2} and their antiparallel diodes D_{A1} and D_{A2}. This allows for ZVS conditions for the main switches. The auxiliary switches are turned on and off under ZCS conditions.

ARCP inverters, typically designed for high power ratings (in excess of 1 MVA), provide highly efficient power conversion. In contrast to the RDCL inverter, whose output voltage waveforms consist of packets of resonant pulses, the ARCP inverter is capable of true pulse width modulation. Typically, IGBTs (insulated-gate bipolar transistors) or GTOs (gate turn-off thyristors) are used as the main switches, while MCTs (MOSFET-controlled thyristors) or IGBTs serve as the auxiliary switches.

4.4 FREQUENCY CHANGERS

A cascade of a rectifier, dc link, and inverter can be thought of as a *frequency changer,* in which the fixed-frequency input voltage and current are converted into adjustable-frequency output variables. Frequency changers with a reversible power flow are of particular interest in the drive technology. The rectifier in such a scheme may be of the phase-controlled or PWM type. The voltage source and current source PWM frequency changers are shown in Figures 4.18 and 4.19, respectively.

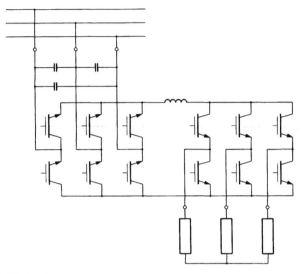

FIGURE 4.18 Voltage source PWM frequency changer.

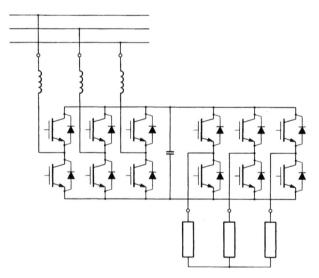

FIGURE 4.19 Current source PWM frequency changer.

They are characterized by high-quality input currents, and they can transfer power in both directions.

For dissipation of electric power sporadically drawn from an induction machine fed by a voltage source inverter with a diode rectifier, a *braking resistor* is used. The braking resistor, in series with a semiconductor power

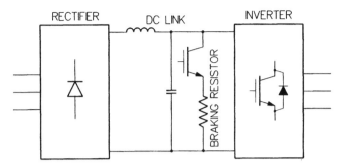

FIGURE 4.20   Braking resistor arrangement.

switch, is inserted between the dc link and inverter, as illustrated in Figure 4.20.

4.5 CONTROL OF VOLTAGE SOURCE INVERTERS

To explain the principles of control of inverters, it is convenient to introduce the so-called *switching variables*, variously defined depending on the type of inverter. For the most common, two-level voltage source inverter depicted in Figure 4.13, three binary switching variables, a, b, and c, one per phase of the inverter, are sufficient. As already mentioned, both switches in an inverter leg cannot be on simultaneously, because they would short the dc supply source (i.e., the dc-link capacitor). The situation when both switches are off is not dangerous, but the voltage at the corresponding output terminal is undetermined. This is so because, depending on the polarity of the load current, the terminal would be connected, via one of the freewheeling diodes, to either the positive or negative dc bus. Therefore, in practice, except for the very short blanking-time intervals, one switch in each phase is on, and the other is off. Consequently, each inverter leg can assume two states only, and the number of states of the whole inverter is eight (2^3).

Taking as an example phase A, the switching variable a is defined to assume the value of 1 if switch SA is on and switch SA$'$ is off. If, conversely, SA is off and SA$'$ is on, a assumes the value of 0. The other two switching variables, b and c, are defined analogously. An inverter state can be denoted as abc_2. For example, with $a = 1$, $b = 0$, and $c = 1$, the inverter is said to be in State 5, because $101_2 = 5$.

It is easy to show that the line-to-line output voltages, v_{AB}, v_{BC}, and v_{CA}, of the voltage source inverter are given by

$$\begin{bmatrix} v_{AB} \\ v_{BC} \\ v_{CA} \end{bmatrix} = V_i \begin{bmatrix} 1 & -1 & 0 \\ 0 & 1 & -1 \\ -1 & 0 & 1 \end{bmatrix} \begin{bmatrix} a \\ b \\ c \end{bmatrix}. \tag{4.3}$$

When the same control principle is applied to all three phases of an inverter feeding a balanced wye-connected load, the individual line-to-neutral voltages, v_{AN}, v_{BN}, and v_{CN}, are balanced too; that is,

$$v_{AN} + v_{BN} + v_{CN} = 0. \tag{4.4}$$

In addition (unconditionally),

$$v_{AN} - v_{BN} = v_{AB} \tag{4.5}$$

and

$$v_{BN} - v_{CN} = v_{BC}. \tag{4.6}$$

Solving Eqs. (4.4) to (4.6) for v_{AN}, v_{BN}, and v_{CN}, yields

$$\begin{bmatrix} v_{AN} \\ v_{BN} \\ v_{CN} \end{bmatrix} = \frac{1}{3} \begin{bmatrix} 1 & 0 & -1 \\ -1 & 1 & 0 \\ 0 & -1 & 1 \end{bmatrix} \begin{bmatrix} v_{AB} \\ v_{BC} \\ v_{CA} \end{bmatrix}, \tag{4.7}$$

which, when combined with Eq. (4.3), gives

$$\begin{bmatrix} v_{AN} \\ v_{BN} \\ v_{CN} \end{bmatrix} = \frac{V_i}{3} \begin{bmatrix} 2 & -1 & -1 \\ -1 & 2 & -1 \\ -1 & -1 & 2 \end{bmatrix} \begin{bmatrix} a \\ b \\ c \end{bmatrix}. \tag{4.8}$$

The simplest control strategy for the inverter consists in imposing the 5-4-6-2-3-1 state sequence, resulting in the already-mentioned square-wave, or *six-step*, mode of operation. Waveforms of the line-to-line and line-to-neutral output voltages of the inverter in this mode are shown in Figure 4.21. Each switch of the inverter is turned on and off once per cycle only, and the peak value of the fundamental line-to-line output voltage is $1.1 V_i$. However, the load current is of poor quality, due to the high content of low-order voltage harmonics. Also, the magnitude of output voltage cannot be controlled within the inverter, which constitutes another disadvantage of the square-wave mode. Therefore, in most practical inverters, transition to the square-wave operation occurs only when

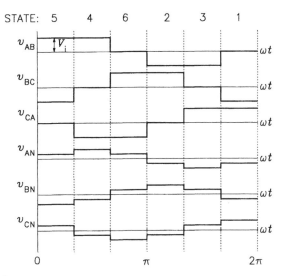

FIGURE 4.21   Output voltage waveforms in a voltage source inverter in the square-wave mode.

the maximum possible output voltage is needed. Other than that, the inverter operates in the PWM mode.

Typical voltage waveforms in a PWM inverter are illustrated in Figure 4.22. In this example, the period of output voltage is divided into 12 so-called *switching intervals*. One pulse of each switching variable appears in each switching interval, and the adjustable pulse width varies from zero to the interval width. The number, N, of switching intervals per cycle of the output voltage, is given by

$$N = \frac{f_{sw}}{f}, \qquad (4.9)$$

where f_{sw} denotes the so-called switching frequency and f is the fundamental output frequency of inverter. The switching frequency is usually constant; thus, N depends on the output frequency only, and it is not necessarily an integer. The voltage waveforms are pulsed, not sinusoidal, and clusters of the most pronounced high harmonics coincide with multiples of the switching frequency. Thus, with typical switching frequencies on the order of several kHz, harmonic currents generated by those harmonics are weak, thanks to the low-pass action of load (motor) inductances. As a result, current waveforms in PWM inverters are close to ideal sinusoids, with only a small ripple.

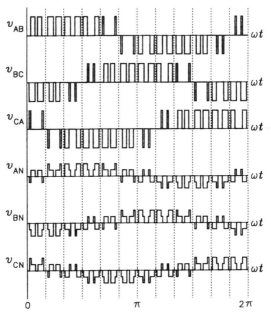

FIGURE 4.22   Output voltage waveforms in a voltage source inverter in the PWM mode.

The most common class of PWM techniques for voltage source inverters is based on the idea of *voltage space vectors*. Space vectors of electric and magnetic variables are an important tool in the analysis, modeling, and control of three-phase ac machines, and their physical interpretation will be given in Chapter 6. Here, only the formal definition of voltage space vectors is provided.

Considering three-phase voltages v_a, v_b, and v_c, the corresponding voltage space vector, v, is given by

$$v = v_d + jv_q, \qquad (4.10)$$

where

$$\begin{bmatrix} v_d \\ v_q \end{bmatrix} = \begin{bmatrix} 1 & -\dfrac{1}{2} & -\dfrac{1}{2} \\[2mm] 0 & \dfrac{\sqrt{3}}{2} & -\dfrac{\sqrt{3}}{2} \end{bmatrix} \begin{bmatrix} v_a \\ v_b \\ v_c \end{bmatrix} \qquad (4.11)$$

Voltages v_a, v_b, and v_c can denote the line-to-neutral voltages, that is, $v_a = v_{AN}$, $v_b = v_{BN}$, $v_c = v_{CN}$, or the line-to-line voltages, that is, $v_a = v_{AB}$,

$v_b = v_{BC}$, $v_c = v_{CA}$. If $v_a + v_b + v_c = 0$, the abc→dq transformation defined by Eq. (4.11) can be inverted, to yield

$$\begin{bmatrix} v_a \\ v_b \\ v_c \end{bmatrix} = \begin{bmatrix} \dfrac{2}{3} & 0 \\ -\dfrac{1}{3} & \dfrac{1}{\sqrt{3}} \\ -\dfrac{1}{3} & -\dfrac{1}{\sqrt{3}} \end{bmatrix} \begin{bmatrix} v_d \\ v_q \end{bmatrix}. \tag{4.12}$$

The abc→dq transformation transforms three actual voltages into a two-dimensional space vector in a complex plane. In particular, considering the classic two-level voltage source inverter, a voltage vector can be assigned to each of the eight states.

EXAMPLE 4.1   To illustrate the idea of voltage space vectors, the space vector, v_5, of the line-to-neutral output voltage of a two-level voltage source inverter in State 5 will be found. The inverter is assumed to be supplied with an ideal dc input voltage, V_i.

In State 5, the switching variables are: $a = 1$, $b = 0$, and $c = 1$. Thus, according to Eq. (4.8),

$$\begin{bmatrix} v_{AN} \\ v_{BN} \\ v_{CN} \end{bmatrix} = \frac{V_i}{3} \begin{bmatrix} 2 & -1 & -1 \\ -1 & 2 & -1 \\ -1 & -1 & 2 \end{bmatrix} \begin{bmatrix} 1 \\ 0 \\ 1 \end{bmatrix} = \frac{V_i}{3} \begin{bmatrix} 1 \\ -2 \\ 1 \end{bmatrix},$$

and, from Eq. (4.11),

$$\begin{bmatrix} v_{d5} \\ v_{q5} \end{bmatrix} = \begin{bmatrix} 1 & -\dfrac{1}{2} & -\dfrac{1}{2} \\ 0 & \dfrac{\sqrt{3}}{2} & -\dfrac{\sqrt{3}}{2} \end{bmatrix} \frac{V_i}{3} \begin{bmatrix} 1 \\ -2 \\ 1 \end{bmatrix} = \frac{V_i}{2} \begin{bmatrix} 1 \\ -\sqrt{3} \end{bmatrix};$$

that is,

$$v_5 = v_{d5} + jv_{q5} = \frac{1}{2}V_i - j\frac{\sqrt{3}}{2}V_i \qquad ■$$

Space vectors of the line-to-neutral voltage associated with the eight states of the voltage source inverters are shown in Figure 4.23 in the per-unit format, with the dc input voltage, V_i, taken as the base voltage. There are six nonzero vectors, v_1 through v_6, and two zero vectors, v_0 and v_7,

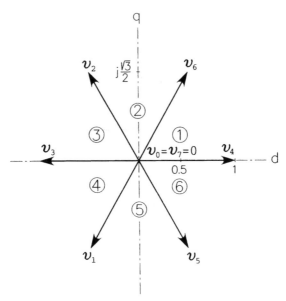

FIGURE 4.23 Space vectors of the line-to-neutral voltage (per-unit) in the two-level voltage source inverter.

resulting from clamping all the output terminals to the negative dc bus (State 0) or positive bus (State 7). The nonzero vectors divide the complex plane into six sectors (sextants), numbered from 1 to 6. Comparing Figures 4.21 and 4.23, we see that the sequence of inverter states corresponding to the consecutive nonzero voltage vectors results in the square-wave mode of operation of the inverter.

It can easily be shown that if voltages v_a, v_b, and v_c form a balanced set of three-phase voltages, the resultant voltage space vector, v, has the magnitude, V, 1.5 times greater than the peak value of those voltages. The phase angle, α, of the voltage vector equals that of v_a. Thus, if

$$v_a(t) = V_m\cos(\omega t), \tag{4.13}$$

where V_p denotes a peak value, then

$$v = Ve^{j\alpha} = 1.5V_m e^{j\omega t}. \tag{4.14}$$

As the time progresses, vector v revolves in the complex plane with the angular velocity ω. Consequently, the goal of inverter control can be formulated as follows: Make the space vector of the output voltage to rotate with a desired speed, and adjust the magnitude of this vector to a desired value.

As shown in Figure 4.23, the voltage source inverter can produce only stationary voltage vectors. The popular Space Vector Pulse Width Modulation (SVPWM) technique overcomes this limitation by means of generation of the stationary vectors in such a manner that it is their *time average* that follows the revolving reference vector, v^*. Specifically, to synthesize the desired voltage vector in a given sextant, the two nonzero stationary vectors framing this sextant, plus a zero vector (or vectors), are used. To explain this technique, the reference vector, at a certain instant, is assumed to lie in Sextant 3, as illustrated in Figure 4.24. The local (within the sextant) angular position of v^*, β, is given by

$$\beta = \alpha - \frac{\pi}{3}int\left(\frac{3}{\pi}\alpha\right), \tag{4.15}$$

and the vector of output voltage is assembled from framing vectors, v_X and v_Y (v_2 and v_3 in Sextant 3) and the zero vector (or vectors), v_Z (v_0 or v_7). Specifically, inverter states producing vectors v_X, v_Y, and v_Z are imposed with durations $d_X T_{sw}$, $d_Y T_{sw}$, and $d_Z T_{sw}$, respectively, where $T_{sw} = 1/f_{sw}$ denotes the switching period, that is, the length of the switching interval. Coefficients d_X, d_Y, and d_Z, which can be called *state duty ratios*, express relative durations of individual states. Clearly,

$$d_X + d_Y + d_Z = 1. \tag{4.16}$$

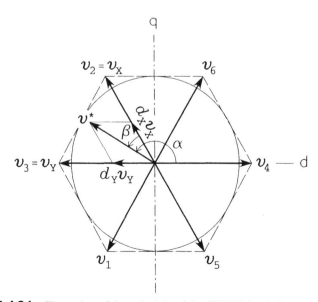

FIGURE 4.24 Illustration of the principle of the SVPWM technique.

Duty ratios d_X and d_Y can be found from equation

$$v^* = d_X v_X + d_Y v_Y \qquad (4.17)$$

decomposed into its real and imaginary parts to produce two real-coefficient equations. The solution is

$$d_X = m\sin\left(\frac{\pi}{3} - \beta\right) \qquad (4.18)$$

and

$$d_Y = m\sin(\beta), \qquad (4.19)$$

where m denotes the so-called *modulation index*. The modulation index can be defined in terms of the magnitude, V, of the voltage vector and that, V_{max}, of the maximum voltage vector possible to be generated using pulse width modulation, as

$$m = \frac{V}{V_{max}}. \qquad (4.20)$$

The duty ratio, d_Z, of the zero state is found from Eq. (4.16) as

$$d_Z = 1 - d_X - d_Y. \qquad (4.21)$$

The maximum voltage vector is obtained when only the nonzero vectors are utilized, that is, $d_Z = 0$. Then, the trajectory of the vector forms the circle shown in Figure 4.24. Consequently (see also Figure 4.23), $V_{max} = \sqrt{3}V_i/2$. It means that the maximum available peak value of the fundamental line-to-neutral output voltage of a PWM inverter is $V_i/\sqrt{3}$ [see Eq. (4.14)], and that of the fundamental line-to-line voltage equals the supply dc voltage, V_i. For comparison, as already mentioned, the fundamental output voltage in the square-wave mode of operation is $1.1\ V_i$.

Eqs. (4.18), (4.19), and (4.21) specify durations of individual states of the inverter but not their sequence within the switching interval. Two such sequences, tentatively called a high-performance sequence and a high-efficiency sequence, are commonly used. The high-performance state sequence is | X - Y - Z$_1$ | Y - X - Z$_2$ | ..., where zero states Z$_1$ and Z$_2$ are such that the transition from one state to another involves switching in one phase only. It means that only one switching variable changes its value from zero to one or vice versa. For example, in Sextant 2, where X = 6 and Y = 2, the high-quality sequence is | 6 - 2 - 0 | 2 - 6 - 7 |

..., that is, in the binary, *abc*, notation, | 110 - 010 - 000 | 010 - 110 - 111 |.... This state sequence, which for a given number, *N*, of switching intervals per cycle of the output voltage results in the best quality of output currents, yields *N*/2 switching pulses per switch and per cycle.

The number of switchings can further be reduced, at the expense of slightly increased distortion of current waveforms, when the high-efficiency state sequence, | X - Y - Z | Z - Y - X | ..., is employed. Here, Z = 0 in even sextants, and Z = 7 in odd sextants. If Sextant 2 is again used as an example, the high-efficiency sequence is | 6 - 2 - 0 | 0 - 2 - 6 | ..., or | 110 - 010 - 000 | 000 - 010 - 110 |.... Note that switching variable *c* is zero throughout the whole sextant in question. With this state sequence, the number of switching pulses per switch and per cycle is *N*/3 + 1, that is, almost one-third lower than that with the high-quality sequence, which results in a proportional reduction of switching losses in the inverter.

EXAMPLE 4.2   A switching interval within which the space vector of output voltage is to follow the reference vector $v^* = 240 \angle 170°$ V is considered. The inverter is supplied from a 430-V dc voltage source, and the switching frequency is 2 kHz. Find the duration and high-quality sequence of inverter states in the interval in question.

The reference voltage vector is in Sextant 3 and, according to Eq. (4.18), the local angular position of v^* is 50°. The maximum available magnitude of the space vector of output voltage is $v_{max} = \sqrt{3} \times 430/2 = 372.4$ V. Thus, the modulation index is $m = 240/372.4 = 0.644$. Consequently, the duty ratios of individual states are $d_X = 0.644 \sin(60° - 50°) = 0.112$, $d_Y = 0.644 \sin(50°) = 0.493$, and $d_Z = 1 - 0.112 - 0.493 = 0.395$. Because the switching interval, T_{sw}, is 0.5 ms long, the State X should last $0.112 \times 0.5 = 0.056$ ms, State Y should last $0.493 \times 0.5 = 0.247$ ms, and State Z should last $0.395 \times 0.5 = 0.197$ ms.

Sextant 3 is framed by vectors v_2 and v_3. Hence, assuming the X - Y - Z_1 state sequence in the considered switching interval, the sequence of states is 2 (0 to 0.056 ms), followed by 3 (0.056 ms to 0.303 ms), and 7 (0.303 ms to 0.5 ms). Note that in the next switching interval the state sequence should be Y - X - Z_2, that is, 3 - 2 - 0 (assuming that the reference vector will still be in the same sextant).   ∎

Typical waveforms of the output current, i_A, in the voltage source inverter in the square-wave and PWM operation modes are shown in Figures 4.25(a) and 4.25(b), respectively, with an induction motor as a

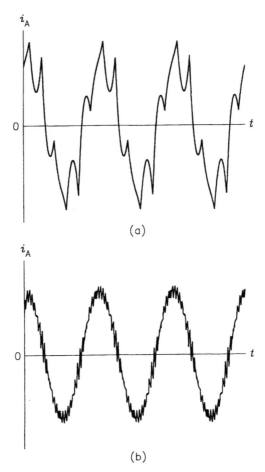

FIGURE 4.25 Waveforms of the output current in a voltage source inverter feeding an induction motor: (a) square-wave operation, (b) PWM operation.

load. Note the similarity of the *output* current in a PWM inverter to that of the *input* current in a PWM rectifier (see Figure 4.10).

In the three-level inverter, three states are employed for each leg, so the switching variables are of the ternary format. Specifically,

$$a = \begin{cases} 0 \;\; \textit{if S1, S2 are OFF and S3, S4 are ON} \\ 1 \;\; \textit{if S1, S4 are OFF and S2, S3 are ON} \\ 2 \;\; \textit{if S1, S2 are ON and S3, S4 are OFF.} \end{cases} \qquad (4.22)$$

The output line-to-line voltages are given by

$$\begin{bmatrix} v_{AB} \\ v_{BC} \\ v_{CA} \end{bmatrix} = \frac{V_i}{2} \begin{bmatrix} 1 & -1 & 0 \\ 0 & 1 & -1 \\ -1 & 0 & 1 \end{bmatrix} \begin{bmatrix} a \\ b \\ c \end{bmatrix}, \qquad (4.23)$$

and the line-to-neutral ones by

$$
\begin{bmatrix} v_{AN} \\ v_{BN} \\ v_{CN} \end{bmatrix} = \frac{V_i}{6} \begin{bmatrix} 2 & -1 & -1 \\ -1 & 2 & -1 \\ -1 & -1 & 2 \end{bmatrix} \begin{bmatrix} a \\ b \\ c \end{bmatrix}. \tag{4.24}
$$

The abc_3 notation can be used for the 27 states of the three-level inverter. Corresponding per-unit space vectors of the line-to-neutral voltage are shown in Figure 4.26. States 0, 13, and 26 produce zero vectors.

The greater number of states allows for higher quality currents than those generated in two-level inverters. Even the square-wave mode of operation, illustrated in Figure 4.27, results in line-to-neutral voltage waveforms well approximating sinusoids. In the PWM mode, to produce currents of comparable quality, a three-level inverter can operate with a switching frequency much lower than that of a two-level inverter.

As shown in Figure 4.28, in the RDCL inverters, individual "pulses" of the output voltage are actually trains of resonant, usually clipped, pulses of this voltage, an integer number of pulses in each train. This somewhat limits the flexibility of control of these inverters. In contrast, as already mentioned, the ARCP inverters are capable of true pulse width modulation, that is, the output voltage pulses are similar to those in hard-switching inverters.

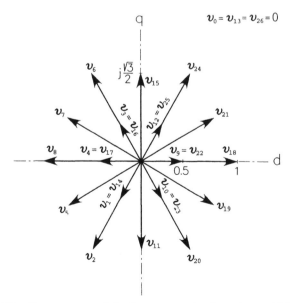

FIGURE 4.26   Space vectors of the line-to-neutral voltage (per unit) in the three-level voltage source inverter.

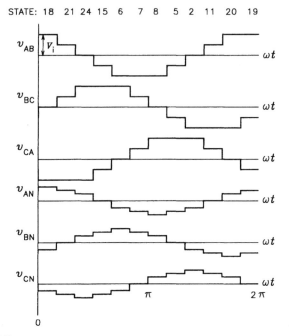

FIGURE 4.27   Output voltage waveforms in a three-level inverter in the square-wave operation mode.

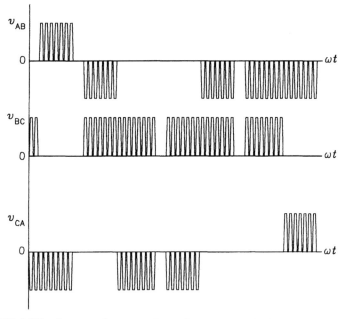

FIGURE 4.28   Output voltage waveforms in a resonant dc link inverter.

In many ASDs, the open-loop voltage control is insufficient to ensure the required quality of dynamic performance of the drive. In these drives, the motor current is closed-loop controlled, the voltage control being subordinated to the current control. Two basic approaches to current control, each with a number of variants, have emerged over the years, the *bang-bang control* and *linear control.*

The simplest version of the bang-bang control, based on the so-called *hysteresis controllers,* is illustrated in Figure 4.29. In each phase, the actual current is compared with the reference current, and the difference (current error) is applied to a hysteresis controller, whose output signal constitutes the switching variable for this phase. Taking phase A as an example, the characteristic of the hysteresis controller is given by

$$
a = \begin{cases} 0 & if \ \Delta i_A < -\dfrac{h}{2} \\[2mm] 1 & if \ \Delta i_A > \dfrac{h}{2} \end{cases}, \tag{4.25}
$$

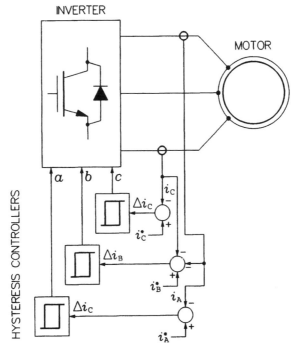

FIGURE 4.29   Block diagram of the bang-bang current control scheme.

where Δi_A denotes the current error and h is the width of the tolerance band. With the current error within the tolerance band, the value of a remains unchanged. The bang-bang current control is characterized by a fast response to rapid changes of the reference current. Many modifications of the basic scheme have been proposed to stabilize the switching frequency and reduce the interaction between phases.

In a linear current control system, linear controllers are used to generate reference signals for the inverter's pulse width modulator. Such a system is shown in Figure 4.30. Output currents i_A and i_C are measured and converted into the i_d and i_q components of the current space vector $\mathbf{i}$. Specifically,

$$\mathbf{i} = i_d + ji_q, \tag{4.26}$$

where

$$i_d = \frac{3}{2}i_A \tag{4.27}$$

and

$$i_q = -\frac{\sqrt{3}}{2}i_A - \sqrt{3}i_C. \tag{4.28}$$

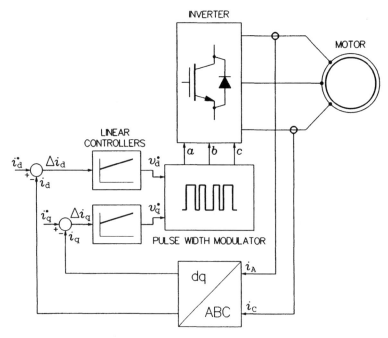

FIGURE 4.30 Block diagram of the linear current control scheme.

Signals i_d and i_q are compared with their reference counterparts, i_d^* and i_q^*, and the respective current errors, Δi_d and Δi_q, are applied to linear controllers, typically of the proportional-integral (PI) type. These produce components v_d^* and v_q^* of the reference voltage vector, v^*, for the pulse width modulator, which generates the switching variables, a, b, and c, for the inverter. Usually, the SVPWM technique is used in the modulator.

4.6 CONTROL OF CURRENT SOURCE INVERTERS

Current source inverters are less commonly used in induction motor ASDs than voltage source inverters, mostly due to the poorer quality of output currents. Still, they have certain advantages, such as imperviousness to overcurrents, even with a short circuit in the inverter or the load. The absence of freewheeling diodes further increases the reliability. Also, current source inverters are characterized by inherently excellent dynamics of the phase angle control of the output current. Such control is realized by changing the state of inverter and the resultant redirecting of the source current. However, the magnitude adjustments of output currents can only be performed in the supplying rectifier. The rectifier allows bidirectional flow of power, and, because the input current is always positive, the input voltage becomes negative when the power flows from the load to the supply power system. Therefore, semiconductor power switches used in a current source inverter must have the reverse blocking capability.

In contrast with voltage source inverters, the simultaneous on-state of both switches in the same inverter leg is safe and recommended for a short period of time initiating a state change of the inverter. This is to avoid the danger of interrupting the current, which would result in an overvoltage. Consequently, switching variables are defined differently than those in the voltage source inverter. In the subsequent considerations, variables a, b, and c are assigned to switches SA, SB, and SC (e.g., $a = 1$ means that SA is on), and variables a', b', and c' to switches SA', SB', and SC' (see Figure 4.14). Then, the output line currents, i_A, i_B, and i_C, of the current source inverter can be expressed as

$$i_A = (a - a')I_i,$$
$$i_B = (b - b')I_i,$$

(4.29)

and

$$i_C = (c - c')I_i,$$

where I_i denotes the constant input current. If the motor has a delta-connected stator, then the currents, i_{AB}, i_{BC}, and i_{CA}, in the individual phase windings are given by

$$\begin{bmatrix} i_{AB} \\ i_{BC} \\ i_{CA} \end{bmatrix} = \frac{1}{3} \begin{bmatrix} 1 & -1 & 0 \\ 0 & 1 & -1 \\ -1 & 0 & 1 \end{bmatrix} \begin{bmatrix} i_A \\ i_B \\ i_C \end{bmatrix}. \qquad (4.30)$$

The peak value of the fundamental line currents is approximately $1.1\ I_i$.

Switching variables in the square-wave operation mode of the current source inverter are shown in Figure 4.31 and the resultant output current waveforms in Figure 4.32. It can be seen that at any time only two switches

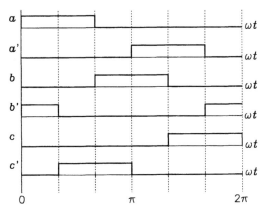

FIGURE 4.31 Switching variables in the current source inverter in the square-wave operation mode.

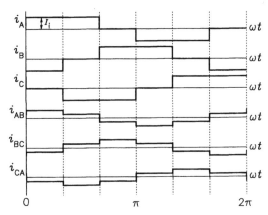

FIGURE 4.32 Output current waveforms in a current source inverter in the square-wave operation mode.

conduct the output currents. This conduction pattern is typical for the voltage source rectifier which, indeed, is an inverse of the current source inverter. Space vectors of the line currents associated with individual states of the inverter are depicted in Figure 4.33 in the per-unit format, with the dc input current, I_i, taken as the base current. The states are designated by letters denoting conducting switches of the inverter. For example, State AB represents the situation in which the conducting switches are SA and SB'. Simultaneous conduction of both switches in the same leg results in a zero vector (States AA, BB, and CC).

Similarly to the voltage source inverter, the square-wave mode of operation requires each switch to be turned on and off once per cycle only. This is one of the reasons that current source inverters have typically been used in high-power drives, with large and slow semiconductor power switches employed in the inverter.

PWM current source inverters, equipped with output capacitors (see Figure 4.14), are characterized by significantly higher quality of output currents than that in square-wave inverters. Several PWM techniques have been developed, one of them, based on a trapezoidal modulating function, illustrated in Figure 4.34. The modulating function signal, x, is compared with a triangular carrier signal, y, and the intersection instants determine

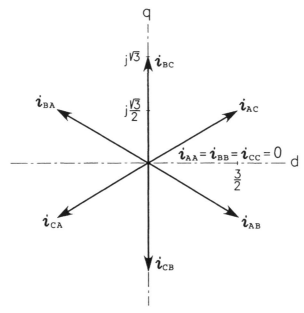

FIGURE 4.33   Space vectors of the line currents (per unit) in the current source inverter.

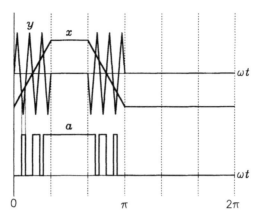

FIGURE 4.34 Illustration of a PWM technique for a current source inverter.

the value of switching variable a. With respect to a, switching variables b and c are delayed by one-third and two-thirds of the period of output current. Variables a', b', and c' are shifted with respect to a, b, and c by a half period. The best attenuation of low-order harmonics in the output currents is achieved with the peak value of the modulating function equal 0.82 of the peak of the carrier signal. PWM current source inverters are not feasible for high-power drives because of the excessive size of the required capacitors.

4.7 SIDE EFFECTS OF CONVERTER OPERATION IN ADJUSTABLE SPEED DRIVES

Although effective and efficient, power electronic converters used in ASDs give birth to certain undesirable side effects. Their extent and severity have spawned the popular phrase about "secondary issues, but primary concerns."

The problem of low-order current harmonics generated by diode rectifiers in the supply system has already been mentioned in Section 4.2. To avoid bulky passive input filters, much smaller active filters or PWM rectifiers can be used instead. However, their mode of operation results in clusters of high-order harmonics in the input currents. These high-frequency currents cause, in turn, high-frequency components of voltage in the utility grid, called *voltage noise* or, more generally, *conducted electromagnetic interference (EMI)*. Also, certain amount of voltage noise leaks to the supply system from PWM inverters via the dc link and input rectifier. In addition, the very process of switching fast semiconductor power switches produces electromagnetic noise in the megahertz range of frequency. Part of the energy carried by high-frequency noise is radiated by overhead transmission lines and other wiring acting as antennas. Because both the conducted and radiated EMI tend to disturb the operation of communication systems and other sensitive electronic equipment, a special low-pass filter, called the EMI or RF (radio frequency) filter, must be installed at the input to the rectifier-inverter cascade.

High values of the rate of voltage change, dv/dt, resulting from the short rise times of voltage pulses generated by the inverter, cause insulation degradation in stator windings due to unequal voltage distribution among individual coils. Another voltage hazard is associated with the cable connecting the motor to the inverter. For voltage pulses, the cable, if sufficiently long, constitutes a transmission line. The ringing and reflected-wave phenomena result in overvoltages reaching twice the normal amplitude of the pulses if the cable exceeds certain critical length (the rule of thumb for the critical length is 60 meters per each microsecond of the rise time). Clearly, the long-cable overvoltages are hazardous for the stator winding, and they must be reduced by resistive-capacitive (RC) filters placed at either end of the cable. The filters, a necessary evil, increase the bulk and cost of the drive system.

The induction motor is a three-wire load, with individual phases coupled to ground via stray capacitances of the motor. The stray capacitances, although small, provide low-impedance paths to ground for transient currents generated by high values of dv/dt of the inverter output voltage. One of these capacitances is that of the motor bearings, in which the thin film of lubricant constitutes a dielectric. Electric charge accumulates on the rotor assembly until the resultant shaft-to-ground voltage exceeds the dielectric capability of the bearing lubricant. The flashover currents significantly accelerate wear of the bearings due to so-called electrical discharge machining (EDM). Transverse grooves, pits, or frosting may appear in the bearing race after just a few months of operation of the drive. Direct preventive measures proposed include the outer-race insulation, conductive lubricant, dielectric-metallic Faraday shield in the motor airgap, or shaft grounding system. To reduce the common-mode (neutral to ground), dv/dt induced leakage current, a common-mode choke or transformer can be added at the motor terminals.

The common-mode voltage, responsible for the shaft voltage and bearing and leakage currents, can be reduced by software and hardware means. Denoting by v_A, v_B, and v_C voltages of stator terminals with respect to ground, the common-mode voltage, v_{cm} is given by

$$v_{cm} = \frac{1}{3}(v_A + v_B + v_C). \tag{4.31}$$

If the dc-link capacitance is composed of two capacitors with a grounded common point, $v_A = V_i(a - 0.5)$, $v_B = V_i(b - 0.5)$, and $v_C = V_i(c - 0.5)$. Consequently, if only inverter voltage vectors v_1, v_3, and v_5 or v_2, v_4, and v_6 are used, the common-mode voltage remains constant, that is, it becomes a dc voltage, at the level of $-V_i/6$ or $V_i/6$, respectively. Such PWM strategy is impractical, but it indicates control possibilities for alleviating the problems associated with the common-mode voltage. Several PWM techniques for common-mode voltage reduction have been developed.

Hardware solutions include extended inverter configurations, such as multilevel, double-bridge, and four-leg topologies. Another option is to connect, in parallel with the standard voltage source inverter, an active (switched) circuit for cancellation of the common-mode voltage. Such a circuit, based on an emitter follower, is shown in Figure 4.35. It imposes, via a transformer coupling, a compensating voltage at the inverter output.

Voltage pulses of fixed frequency result in annoying tonal noise in motors fed from PWM inverters. To quiet a drive system, the switching frequency should be raised to a supersonic (greater than 20 kHz) level. This is not always feasible, and random PWM can be used instead. It consists in random changes of the switching frequency, so that the harmonic power (watts) in the voltage spectrum is transferred to continuous power density (watts per hertz). As a result, the tonal noise is converted into "static," easily blending with the background noise. The random

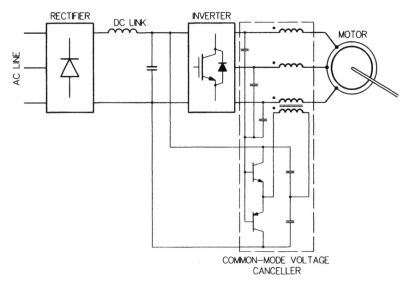

COMMON-MODE VOLTAGE
CANCELLER

FIGURE 4.35 Common-mode voltage canceler for a voltage source inverter.

PWM also reduces the peak and quasi-peak EMI conducted from the drive system to the utility grid.

4.8 SUMMARY

Induction motors in ASDs are supplied from inverters, which are dc to ac power electronic converters. The dc supply voltage for inverters is provided by rectifiers. These usually have the three-phase bridge topology and are based on power diodes or, if control of the input current or inverted power flow is required, on SCRs. Poor quality of currents drawn from the power system is the major disadvantage of such rectifiers, and expensive input filters must be installed to alleviate this problem. The size of input filters can greatly be reduced when PWM rectifiers are employed. They can be of the voltage source or current source type, the latter converters being capable of boosting the output voltage above the peak value of the ac input (line-to-line) voltage.

Two-level voltage source inverters are most common in practice, but alternative topologies, such as soft-switching and three-level inverters, are gaining ground. Current source inverters are also used, usually in high-power drives. These inverters are more robust than voltage source inverters, and the dynamics of current phase control is excellent. The current quality in the square-wave mode of operation is poor, though. In that respect, PWM current source inverters, with output capacitors, are a better solution, but the required size of the capacitors limits the power range of these converters.

A cascade of the rectifier, dc link, and inverter constitutes a frequency changer. PWM frequency changers, characterized by high-performance operation and capable of the reversed power flow, can be of the voltage-source or current-source type. The voltage source PWM frequency changer is based on a voltage source PWM rectifier and a current source inverter. Conversely, in the current source PWM frequency changer, a current source rectifier and a voltage source inverter are employed.

The most popular PWM technique for voltage source inverters is based on the concept of voltage space vectors. A voltage space vector is obtained by transformation of values of actual voltages in a three-phase system into a complex number. Three-phase currents in a three-wire system can similarly be transformed.

In contrast to voltage source inverters which, in the PWM mode, allow control of both the frequency and magnitude of the output voltages, current source inverters are incapable of the magnitude control of output currents. This can only be done in the current source (controlled rectifier) supplying the inverter.

Power electronic converters used in ASDs with induction motors cause serious side effects. These include harmonic pollution of the supplying power system, conducted and radiated electromagnetic interference, insulation degradation in stator windings, overvoltages in the cable connecting the inverter and motor, common-mode voltage resulting in accelerated bearing deterioration and leakage currents, and annoying acoustic noise. Most of these side effects can be remedied using appropriate filters. Other measures, such as multilevel, double-bridge, and four-leg inverters, active circuits for the common-mode voltage cancellation, and modified PWM strategies are also employed in practical drive systems.

5

SCALAR CONTROL METHODS

This chapter introduces two-inductance, Γ and Γ', per-phase equivalent circuits of the induction motor for explanation of the scalar control methods. The open-loop, Constant Volts Hertz, and closed-loop speed control methods are presented, and field weakening and compensation of slip and stator voltage drop are explained. Finally, scalar torque control, based on decomposition of the stator current into the flux-producing and torque-producing components, is described.

5.1 TWO-INDUCTANCE EQUIVALENT CIRCUITS OF THE INDUCTION MOTOR

As a background for scalar control methods, it is convenient to use a pair of two-inductance per-phase equivalent circuits of the induction motor. They differ from the three-inductance circuit introduced in Section 2.3, which can be called a *T-model* because of the configuration of inductances (see Figure 2.14). Introducing the transformation coefficient γ given by

$$\gamma = \frac{X_s}{X_m},$$ (5.1)

the T-model of the induction motor can be transformed into the so-called Γ-*model*, shown in Figure 5.1. Components and variables of this equivalent circuit are related to those of the T-model as follows:

1. Rotor resistance (referred to stator),

$$R_R = \gamma^2 R_r \tag{5.2}$$

2. Magnetizing reactance,

$$X_M = \gamma X_m = X_s. \tag{5.3}$$

3. Total leakage reactance,

$$X_L = \gamma X_{ls} + \gamma^2 X_{lr}. \tag{5.4}$$

4. Rotor current referred to stator,

$$I_R = \frac{I_r}{\gamma}. \tag{5.5}$$

5. Rotor flux,

$$\Lambda_R = \gamma \Lambda_r \tag{5.6}$$

The actual radian frequency, ω_r, of currents in the rotor the induction motor is given by

$$\omega_r = s\omega. \tag{5.7}$$

This frequency, subsequently called *rotor frequency*, is proportional to the slip velocity, ω_{sl}, as

$$\omega_r = p_p \omega_{sl}. \tag{5.8}$$

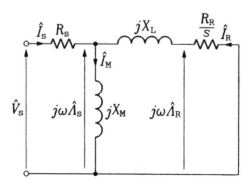

FIGURE 5.1   The Γ equivalent circuit of the induction motor.

Taking into account that $R_R/s = R_R\omega/\omega_r$, current I_R in the Γ-model can be expressed as

$$I_R = |\hat{I}_R| = \frac{\Lambda_s}{R_R} \frac{\omega_r}{\sqrt{(\tau_\Gamma\omega_r)^2 + 1}}, \tag{5.9}$$

where $\tau_\Gamma = L_L/R_R$. Symbol L_L denotes the total leakage inductance ($L_L = X_L/\omega$). The electrical power, P_{elec}, consumed by the motor is

$$P_{elec} = 3R_R\frac{\omega}{\omega_r}I_R^2, \tag{5.10}$$

and the mechanical power, P_{mech}, can be obtained from P_{elec} by subtracting the resistive losses, $3R_R I_R^2$. Finally, the torque, T_M, developed in the motor can be calculated as the ratio of P_{mech} to the rotor angular velocity, ω_M, which is given by

$$\omega_M = \frac{\omega - \omega_r}{p_p}. \tag{5.11}$$

The resultant formula for the developed torque is

$$T_M = 3p_p\frac{\Lambda_s^2}{R_R} \frac{\omega_r}{(\tau_\Gamma\omega_r)^2 + 1}. \tag{5.12}$$

EXAMPLE 5.1 For the example motor, find parameters of the Γ-model and the developed torque.

The Γ-model parameters are: $\gamma = 1.0339$, $R_R = 0.1668$ Ω/ph, $L_M = 0.0424$ H/ph, and $L_L = 0.00223$ H/ph. This yields $\tau_\Gamma = 0.00223/0.1668 = 0.0134$ s. The rms value, Λ_s, of the stator flux under rated operating conditions of the motor, calculated from the T-model in Figure 2.14, is 0.5827 Wb/ph. Under the same conditions, the slip of 0.027 results in the rotor frequency, ω_r, of $0.027 \times 377 = 10.18$ rad/s. These data allow calculation of the rotor current using Eq. (5.9) as $I_R = 0.5827/0.1668 \times 10.18/[(0.0134 \times 10.18)^2 + 1]^{1/2} = 35.24$ A/ph, which corresponds to $I_r = 1.0339 \times 35.24 = 36.44$ A/ph. The developed torque, T_M, can be found from Eq. (5.12) as $T_M = 3 \times 3 \times 0.5827^2/0.1668 \times 10.18/[(0.0134 \times 10.18)^2 + 1] = 183.1$ Nm. These values can be verified using the program used in Example 2.1. ■

Another two-inductance per-phase equivalent circuit of the induction motor, called an *inverse-Γ* or *Γ'-model*, is shown in Figure 5.2. The

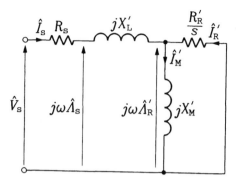

FIGURE 5.2 The Γ' equivalent circuit of the induction motor.

coefficient, γ', for transformation of the T-model into the Γ'-model is given by

$$\gamma' = \frac{X_m}{X_r}, \tag{5.13}$$

and the rotor resistance, magnetizing reactance, total leakage reactance, rotor current, and rotor flux in the latter model are

$$R'_R = \gamma'^2 R_r, \tag{5.14}$$

$$X'_M = \gamma' X_m, \tag{5.15}$$

$$X'_L = X_{ls} + \gamma' X_{lr}, \tag{5.16}$$

$$I'_R = \frac{I_r}{\gamma'}, \tag{5.17}$$

and

$$\Lambda'_R = \gamma' \Lambda_r, \tag{5.18}$$

respectively.

The electrical power is given by an equation similar to Eq. (5.10), that is,

$$P_{elec} = 3R'_R \frac{\omega}{\omega_r} I'^2_R, \tag{5.19}$$

and the developed torque can again be calculated by subtracting the resistive losses and dividing the resultant mechanical power by the rotor velocity. This yields

$$T_M = 3p_p R'_R \frac{I'^2_R}{\omega_r}, \tag{5.20}$$

which, based on the Γ' equivalent circuit, can be rearranged to

$$T_M = 3p_p L'_M I'_R I'_M, \tag{5.21}$$

where $L'_M = X'_M/\omega$ denotes the magnetizing inductance in the Γ'-model.

EXAMPLE 5.2   Repeat Example 5.1 for the Γ'-model of the example motor.

The Γ'-model parameters are: $\gamma' = 0.9823$, $R'_R = 0.1505$ Ω/ph, $L'_M = 0.0403$ H/ph, and $L'_L = 0.0021$ H/ph. Phasors of the rated stator and rotor current calculated from the T-model (see Example 2.1) are $\hat{I}_s = 35.35 - j17.66$ A/ph and $\hat{I}_r = -36.21 + j4.06$ A/ph, respectively. Hence, $\hat{I}_R = (-36.21 + j4.06)/0.9823 = -36.86 + j4.13$ A/ph and $\hat{I}'_M = \hat{I}_s + \hat{I}_R = -1.51 - j13.6$ A/ph. The magnitudes, I_R and I'_M, of these currents are 37.09 A and 13.68 A, respectively, which yields $T_M = 3 \times 3 \times 0.0403 \times 37.09 \times 13.68 = 184.03$ Nm, a result practically the same as that obtained in Example 5.1.   ∎

5.2 OPEN-LOOP SCALAR SPEED CONTROL (CONSTANT VOLTS/HERTZ)

Analysis of Eq. (5.12) leads to the following conclusions:

1. If $\omega_r = 1/\tau_r$, then the maximum (pull-out) torque, $T_{M,max}$, is developed in the motor. It is given by

$$T_{M,max} = 1.5 p_p \frac{\Lambda_s^2}{L_L}, \tag{5.22}$$

and the corresponding critical slip, s_{cr}, is

$$s_{cr} = \frac{1}{\tau_\Gamma \omega}. \tag{5.23}$$

2. Typically, induction motors operate well below the critical slip, so that $\omega_r \ll 1/\tau_\Gamma$. Then, $(\tau_\Gamma \omega_r)^2 + 1 \approx 1$, and the torque is practically proportional to ω_r. For a stiff mechanical characteristic

of the motor, possibly high flux and low rotor resistance are required.

3. When the stator flux is kept constant, the developed torque is independent of the supply frequency, f. On the other hand, the speed of the motor strongly depends on f [see Eq. (3.3)].

It must be stressed that Eq. (5.12) is only valid when the stator flux is kept constant, independently of the slip. In practice, it is usually the stator voltage that is constant, at least when the supply frequency does not change. Then, the stator flux does depend on slip, and the critical slip is different from that given by Eq. (5.23). Generally, for a given supply frequency, the mechanical characteristic of an induction motor strongly depends on which motor variable is kept constant.

EXAMPLE 5.3   Find the pull-out torque and critical slip of the example motor if the stator flux is maintained at the rated level of 0.5827 Wb (see Example 5.1).

The pull-out torque, $T_{\text{M,max}}$, is $1.5 \times 3 \times 0.5827^2/0.00223 = 685.2$ Nm, and the corresponding critical slip, s_{cr}, at the supply frequency of 60 Hz is $1/(377 \times 0.0134) = 0.198$. Note that these values differ from those in Table 2.2, which were computed for a constant stator voltage.   ■

Assuming that the voltage drop across the stator resistance is small in comparison with the stator voltage, the stator flux can be expressed as

$$\Lambda_s \approx \frac{V_s}{\omega} = \frac{1}{2\pi} \frac{V_s}{f}. \tag{5.24}$$

Thus, to maintain the flux at a constant, typically rated level, the stator voltage should be adjusted in proportion to the supply frequency. This is the simplest approach to the speed control of induction motors, referred to as *Constant Volts/Hertz* (CVH) method. It can be seen that no feedback is inherently required, although in most practical systems the stator current is measured, and provisions are made to avoid overloads.

For the low-speed operation, the voltage drop across the stator resistance must be taken into account in maintaining constant flux, and the stator voltage must be appropriately boosted. Conversely, at speeds exceeding that corresponding to the rated frequency, f_{rat}, the CVH condition cannot be satisfied because it would mean an overvoltage. Therefore, the stator voltage is adjusted in accordance to the following rule:

$$V_s = \begin{cases} (V_{\text{s,rat}} - V_{\text{s,0}})\dfrac{f}{f_{\text{rat}}} + V_{\text{s,0}} & \text{for } f < f_{\text{rat}}, \\ V_{\text{s,rat}} & \text{for } f \geq f_{\text{rat}} \end{cases} \tag{5.25}$$

where $V_{s,0}$ denotes the rms value of the stator voltage at zero frequency. Relation (5.25) is illustrated in Figure 5.3. For the example motor, $V_{s,0}$ = 40 V. With the stator voltage so controlled, its mechanical characteristics for various values of the supply frequency are depicted in Figure 5.4.

Frequencies higher than the rated (base) frequency result in reduction of the developed torque. This is caused by the reduced magnetizing current, that is, a weakened magnetic field in the motor. Accordingly, the motor is said to operate in the *field weakening* mode. The region to the right from the rated frequency is often called the *constant power area*, as distinguished from the *constant torque area* to the left from the said frequency. Indeed, with the torque decreasing when the motor speed increases, the product of these two variables remains constant. Note that the described characteristics of the motor can easily be explained by the

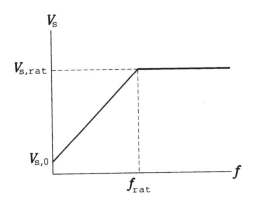

FIGURE 5.3 Voltage versus frequency relation in the CVH drives.

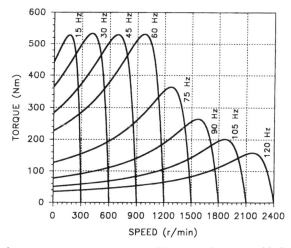

FIGURE 5.4 Mechanical characteristics of the example motor with the CVH control.

impossibility of sustained operation of an electric machine with the output power higher than rated.

A simple version of the CVH drive is shown in Figure 5.5. A fixed value of slip velocity, ω_{sl}, corresponding to, for instance, 50% of the rated torque, is added to the reference velocity, ω_M^*, of the motor to result in the reference synchronous frequency, ω_{syn}^*. This frequency is next multiplied by the number of pole pairs, p_p, to obtain the reference output frequency, ω^*, of the inverter, and it is also used as the input signal to a voltage controller. The controller generates the reference signal, V^*, of the inverter's fundamental output voltage. Optionally, a current limiter can be employed to reduce the output voltage of the inverter when too high a motor current is detected. The current, i_{dc}, measured in the dc link is a dc current, more convenient as a feedback signal than the actual ac motor current.

Clearly, highly accurate speed control is not possible, because the actual slip varies with the load of the motor. Yet, in many practical applications, such as pumps, fans, mixers, or grinders, high control accuracy is unnecessary. The basic CVH scheme in Figure 5.5 can be improved by adding slip compensation based on the measured dc-link current. The ω_{sl} signal is generated in the slip compensator as a variable proportional to i_{dc}. A so modified drive system is shown in Figure 5.6.

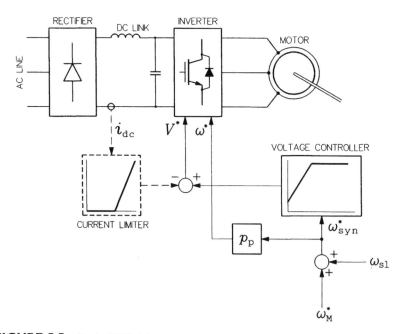

FIGURE 5.5 Basic CVH drive system.

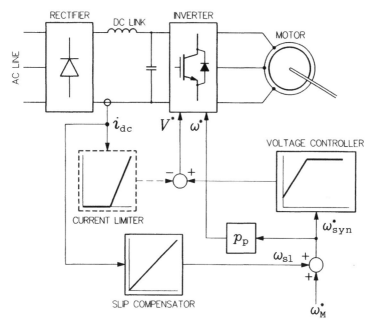

FIGURE 5.6   CVH drive system with slip compensation.

5.3 CLOSED-LOOP SCALAR SPEED CONTROL

With the motor speed measured or estimated, it can be controlled in the closed-loop scheme shown in Figure 5.7. The speed (angular velocity), ω_M, is compared with the reference speed, ω_M^*. The speed error signal, $\Delta\omega_M$, is applied to a slip controller, usually of the PI (proportional-integral) type, which generates the reference slip speed, ω_{sl}^*. The slip speed must be limited for stability and overcurrent prevention. Therefore, the slip controller's static characteristic exhibits saturation at a level somewhat lower than the critical slip speed. When ω_{sl}^* is added to ω_M, the reference synchronous speed, ω_{syn}^*, is obtained. As in the CVH drives in Figures 5.5 and 5.6, the latter signal used to generate the reference values, ω^* and V^*, of the inverter frequency and voltage.

In conjunction with the widespread application of the space vector PWM techniques described in Section 4.5, it is the reference vector, $v^* = V^* e^{j\beta^*}$, of the inverter output voltage that is often produced by the control system. Strictly speaking, the control system determines the reference values m^* of the modulation index and β^* of the voltage vector angle, because these two variables are needed for calculation of duty ratios of inverter states within a given switching interval. Clearly, values

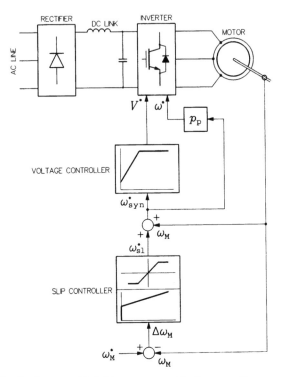

FIGURE 5.7 Scalar-controlled drive system with slip controller.

of m^* and β^* are closely related to those of V^* and ω^*, because $m^* = V^*/V_{max}$ and β^* represents the time integral of ω^*.

5.4 SCALAR TORQUE CONTROL

Closed-loop torque control is typical for winder drives, which are very common in the textile, paper, steel, plastic, or rubber manufacturing industries. In such a drive, one motor imposes the speed while the other provides a controlled braking torque to run the wound tape with constant linear speed and tension. An internal torque-control loop is also used in single-motor ASDs with the closed-loop speed control to improve the dynamics of the drive. Separately excited dc motors, in which the developed torque is proportional to the armature current while the magnetic flux is produced by the field current, are very well suited for that purpose. However, dc motors are more expensive and less robust than the induction ones.

Eq. (5.21) offers a solution for independent control of the flux and torque in the induction motor so that in the steady state it can emulate the separately excited dc motor. It follows from the Γ' equivalent circuit in Fig. 5.2 that the rotor flux can be controlled by adjusting I'_M. On the other hand, with I'_M constant, the developed torque is proportional to I'_R. Because $\hat{I}_s = \hat{I}'_M - \hat{I}'_R$, the stator current can be thought of as a sum of a *flux-producing current*, $\hat{I}_\Phi = \hat{I}'_M$ and a *torque-producing current*, $\hat{I}_T = -\hat{I}'_R$. The question is how to control these two currents by adjusting the magnitude and frequency of stator current.

Assuming that reference values, $\hat{I}^*_\Phi$ and $\hat{I}^*_M$, of the flux-producing and torque-producing components of the stator current are known, the reference magnitude, $\hat{I}^*_s$, of this current is given by

$$I^*_s = \sqrt{I^{*2}_\Phi + I^{*2}_M}.$$ (5.26)

To determine the reference frequency, ω^*_r, of stator current, Eq. (5.21) can be divided by Eq. (5.22) side by side and rearranged to

$$\omega^*_r = \frac{R'_R}{L'_M}\frac{I^*_T}{I^*_\Phi},$$ (5.27)

where ω^*_r denotes the reference rotor frequency. Because

$$\frac{L'_M}{R'_R} = \frac{L_r}{R_r} = \tau_r,$$ (5.28)

where τ_r is the rotor time constant, the reference stator frequency is

$$\omega^* = \omega_o + \omega^*_r = p_p\omega_M + \frac{1}{\tau_r}\frac{I^*_T}{I^*_\Phi},$$ (5.29)

where $\omega_o = \omega - \omega_r = p_p\omega_M$ denotes the rotor velocity of a hypothetical 2-pole motor having the same equivalent circuit as the given $2p_p$-pole motor. The equivalent 2-pole motor is convenient for the analysis and control purposes. Frequencies and angular velocities in both the original, $2p_p$-pole motor and its 2-pole equivalent are compared in Table 5.1.

TABLE 5.1 Frequencies and Angular Velocities in the Actual and Equivalent Motors

Variable	$2p_p$-Pole Actual Motor	2-pole Equivalent Motor
Synchronous frequency/ velocity (rad/s)	$\omega_{syn} = \omega/p_p$	$\omega = p_p\omega_{syn}$
Rotor velocity (rad/s)	$\omega_M = \omega_o/p_p$	$\omega_o = p_p\omega_M$
Slip frequency/velocity (rad/s)	$\omega_{sl} = \omega_r/p_p$	$\omega_r = p_p\omega_{sl}$

A block diagram of the scalar torque control scheme is shown in Figure 5.8. The reference flux-producing and torque-producing currents are computed as

$$I_\Phi^* = \frac{\Lambda_R'^*}{L_M'} \tag{5.30}$$

and

$$I_T^* = \frac{T_M^*}{3p_p\Lambda_R'^*}, \tag{5.31}$$

where $\Lambda_R'^*$ and T_M^* denote reference values of the flux and torque. The former quantity, specific for the Γ' equivalent circuit, is related to the actual rotor flux by Eq. (5.18). The rotor speed, ω_M, appearing in Eq. (5.29), can be measured directly or estimated from other motor variables. The current-controlled inverter must be equipped with current feedback, provided by two current sensors in the output lines.

EXAMPLE 5.4   Determine the reference stator frequency for the example motor if it is to run with the speed of 1500 r/min (157.1 rad/s) torque of 150 Nm, and with the constant rotor flux (referred to Γ'-model), $\Lambda_R'^*$, of 0.5 Wb.

The reference flux- and torque-producing components of the stator current are $I_\Phi^* = 0.5/0.0403 = 12.4$ A/ph (see Example 5.2), and

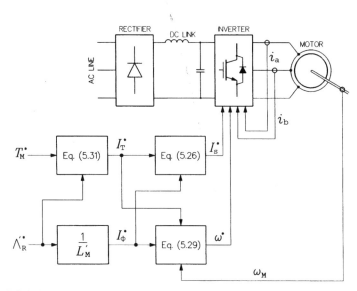

FIGURE 5.8   Scalar torque control scheme.

$I_T^* = 150/(3 \times 3 \times 0.5) = 33.3$ A/ph. The reference stator current, I_s^*, is $(12.4^2 + 33.3^2)^{1/2} = 35.5$ A/ph. The rotor time constant, τ_r, is $0.0417/0.156 = 0.267$ s, and the reference stator radian frequency, ω^*, is $3 \times 157.1 + 1/0.267 \times 33.3/12.4 = 481.4$ rad/s, which represents a frequency of 76.6 Hz.   ■

It must be emphasized that in today's practice of induction motor ASDs, the scalar control methods outlined in Sections 5.3 and 5.4 are considered obsolete. They have only been described as a background for modern vector control methods, which result in much better dynamic performance. If the information about the speed of the motor is available, advanced vector control algorithms can easily be implemented in high-speed digital processors. On the other hand, the CVH drives (also called Volts/Hertz or, simply, variable frequency drives) are still very popular and widely used in low-performance applications.

5.5 SUMMARY

The Γ and Γ' two-inductance steady-state equivalent circuits of the induction motor facilitate explanation of scalar speed and torque control methods. The scalar control, consisting in adjusting the magnitude and frequency of stator voltages or currents, does not guarantee good dynamic performance of the drive, because transient states of the motor are not considered in control algorithms.

In many practical applications, such performance is unimportant, and the CVH drives, with open-loop speed control, are quite sufficient. In these drives, the stator voltage is adjusted in proportion to the supply frequency, except for low and above-base speeds. The voltage drop across stator resistance must be taken into account for low-frequency operation, while with frequencies higher than rated, a constant voltage to frequency ratio would result in overvoltage. Therefore, above the base speed, the voltage is maintained at the rated level, and the magnetic field and maximum available torque decrease with the increasing frequency. Operation of the CVH drives can be enhanced by slip compensation.

Scalar torque control is based on decomposition of the stator current into the flux-producing and torque-producing components. Scalar control techniques with the speed feedback are being phased out by the more effective vector control methods.

6

DYNAMIC MODEL OF THE INDUCTION MOTOR

In Chapter 6, we define and illustrate space vectors of induction motor variables in the stator reference frame, dq. Dynamic equations of the induction motor are expressed in this frame. The idea of a revolving reference frame, DQ, is introduced to transform the ac components of the vectors in the stator frame into dc signals, and formulas for the straight and inverse abc→dq and dq→DQ transformations are provided. We finish by explaining adaptation of dynamic equations of the motor to a revolving reference frame.

6.1 SPACE VECTORS OF MOTOR VARIABLES

Space vectors of three-phase variables, such as the voltage, current, or flux, are very convenient for the analysis and control of induction motors. Voltage space vectors of the voltage source inverter have already been formally introduced in Section 4.5. Here, the physical background of the concept of space vectors is illustrated.

Space vectors of stator MMFs in a two-pole motor have been shown in Chapter 2 in Figures 2.6 through 2.9. The vector of total stator MMF, $\mathscr{F}_s$, is a vectorial sum of phase MMFs, $\mathscr{F}_{as}$, $\mathscr{F}_{bs}$, and $\mathscr{F}_{cs}$, that is,

$$\mathscr{F}_s = \mathscr{F}_{as} + \mathscr{F}_{bs} + \mathscr{F}_{cs} = \mathscr{F}_{as} + \mathscr{F}_{bs}e^{j\frac{2}{3}\pi} + \mathscr{F}_{cs}e^{j\frac{4}{3}\pi}, \qquad (6.1)$$

where $\mathscr{F}_{as}$, $\mathscr{F}_{bs}$, and $\mathscr{F}_{cs}$ denote magnitudes of $\mathscr{F}_{as}$, $\mathscr{F}_{bs}$, and $\mathscr{F}_{cs}$, respectively. In the stationary set of stator coordinates, dq, the vector of stator MMF can be expressed as a complex variable, $\mathscr{F}_s = \mathscr{F}_{ds} + j\mathscr{F}_{qs} = \mathscr{F}_s e^{j\Theta s}$, as depicted in Figure 6.1. Because

$$e^{j\frac{2}{3}\pi} = -\frac{1}{2} + j\frac{\sqrt{3}}{2} \qquad (6.2)$$

and

$$e^{j\frac{4}{3}\pi} = -\frac{1}{2} - j\frac{\sqrt{3}}{2}, \qquad (6.3)$$

then, Eq. (6.1) can be rewritten as

$$\mathscr{F}_s = \mathscr{F}_{ds} + j\mathscr{F}_{qs} = \mathscr{F}_{as} - \frac{1}{2}\mathscr{F}_{bs} - \frac{1}{2}\mathscr{F}_{cs} + j\left(\frac{\sqrt{3}}{2}\mathscr{F}_{bs} - \frac{\sqrt{3}}{2}\mathscr{F}_{cs}\right), \tag{6.4}$$

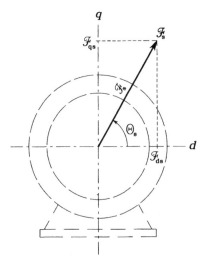

FIGURE 6.1   Space vector of stator MMF.

which explains the abc→dq transformation described by Eq. (4.11). For the stator MMFs,

$$
\begin{bmatrix} \mathscr{F}_{ds} \\ \mathscr{F}_{qs} \end{bmatrix} = \begin{bmatrix} 1 & -\dfrac{1}{2} & -\dfrac{1}{2} \\ 0 & \dfrac{\sqrt{3}}{2} & -\dfrac{\sqrt{3}}{2} \end{bmatrix} \begin{bmatrix} \mathscr{F}_{as} \\ \mathscr{F}_{bs} \\ \mathscr{F}_{cs} \end{bmatrix}
\tag{6.5}
$$

and

$$
\begin{bmatrix} \mathscr{F}_{as} \\ \mathscr{F}_{bs} \\ \mathscr{F}_{cs} \end{bmatrix} = \begin{bmatrix} \dfrac{2}{3} & 0 \\ -\dfrac{1}{3} & \dfrac{1}{\sqrt{3}} \\ -\dfrac{1}{3} & -\dfrac{1}{\sqrt{3}} \end{bmatrix} \begin{bmatrix} \mathscr{F}_{ds} \\ \mathscr{F}_{qs} \end{bmatrix}.
\tag{6.6}
$$

Transformation equations (6.5) and (6.6) apply to all three-phase variables of the induction motor (generally, of any three-phase system), which add up to zero.

Stator MMFs are true (physical) vectors, because their direction and polarity in the real space of the motor can easily be ascertained. Because an MMF is a product of the current in a coil and the number of turns of the coil, the *stator current vector*, i_s, can be obtained by dividing $\mathscr{F}_s$ by the number of turns in a phase of the stator winding. This is tantamount to applying the abc→dq transformation to currents, i_{as}, i_{bs}, and i_{cs} in individual phase windings of the stator. The *stator voltage vector*, v_s, is obtained using the same transformation to stator phase voltages, v_{as}, v_{bs}, and v_{cs}. It can be argued to which extent i_s and v_s are true vectors, but from the viewpoint of analysis and control of induction motors this issue is irrelevant.

It must be mentioned that the abc→dq and dq→abc transformation matrices in Eqs. (6.5) and (6.6) are not the only ones encountered in the literature. As seen in Figure 2.6, when the stator phase MMFs are balanced, the magnitude, $\mathscr{F}_s$, of the space vector, $\mathscr{F}_s$, of the stator MMF is 1.5 times higher than the magnitude (peak value), $\mathscr{F}_{as}$, of phase MMFs. This coefficient applies to all other space vectors. In some publications, the abc→dq transformation matrix in Eq. (6.5) appears multiplied by 2/3, and the dq→abc transformation matrix in Eq. (6.6), by 3/2. Then, the vector magnitude equals the peak value of the corresponding phase quantities. On the other hand, if the product of magnitudes, V_s, and I_s, of stator voltage and current vectors, v_s and i_s, is to equal the apparent power supplied to the stator, the matrices in Eqs. (6.5) and (6.6) should be multiplied by $\sqrt{(2/3)}$ and $\sqrt{(3/2)}$, respectively.

In practical ASDs, the voltage feedback, if needed, is usually obtained from a voltage sensor, which, placed at the dc input to the inverter, measures the dc-link voltage, v_i. The line-to-line and line-to-neutral stator voltages are determined on the basis of current values, a, b, and c, of switching variables of the inverter using Eqs. (4.3) and (4.8). Depending on whether the phase windings of the stator are connected in delta or wye, the stator voltages v_{as}, v_{bs}, and v_{cs} constitute the respective line-to-line or line-to-neutral voltages. Specifically, in a delta-connected stator, $v_{as} = v_{AB}$, $v_{bs} = v_{BC}$, and $v_{cs} = v_{CA}$, while in a wye-connected one, $v_{as} = v_{AN}$, $v_{bs} = v_{BN}$, and $v_{cs} = v_{CN}$.

The current feedback is typically provided by two current sensors in the output lines of the inverter as shown, for instance, in Figure 5.8. The sensors measure currents i_A and i_C, and if the stator is connected in wye, its phase currents are easily determined as $i_{as} = i_A$, $i_{bs} = -i_A - i_C$, and $i_{cs} = i_C$. Because of the symmetry of all three phases of the motor and symmetry of control of all phases of the inverter, the phase stator currents in a delta-connected motor can be assumed to add up to zero. Consequently, they can be found as $i_{as} = (2i_A + i_C)/3$, $i_{bs} = (-i_A - 2i_C)/3$, and $i_{cs} = (-i_A + i_C)/3$. Voltages and currents in the wye- and delta-connected stators are shown in Figure 6.2.

In addition to the already-mentioned space vectors of the stator voltage, $\boldsymbol{v}_s$, and current, $\boldsymbol{i}_s$, four other three-phase variables of the induction motor will be expressed as space vectors. These are the *rotor current vector*, $\boldsymbol{i}_r$, and three *flux-linkage vectors*, commonly, albeit imprecisely, called *flux vectors*: stator flux vector, $\boldsymbol{\lambda}_s$, air-gap flux vector, $\boldsymbol{\lambda}_m$, and rotor flux vector, $\boldsymbol{\lambda}_r$. The air-gap flux is smaller than the stator flux by only the small amount of leakage flux in the stator and, similarly, the rotor flux is only slightly reduced with respect to the air-gap flux, due to flux leakage in the rotor.

EXAMPLE 6.1   To illustrate the concept of space vectors and the static, abc→dq, transformation, consider the example motor operating under rated conditions, with phasors of the stator and rotor current equal $\hat{I}_s = 39.5 \angle -26.5°$ A/ph and $\hat{I}_r = 36.4 \angle 173.6°$ A/ph, respectively. In power engineering, phasors represent rms quantities; thus, assuming that the current phasors in question pertain to phase A of the motor, individual stator and rotor currents are: $i_{as} = \sqrt{2} \times 39.5 \cos(377t - 26.5°) = 55.9 \cos(377t - 26.5°)$ A; $i_{bs} = 55.9 \cos(377t - 146.5°)$ A; $i_{cs} = 55.9 \cos(377t - 266.5°)$ A; $i_{ar} = \sqrt{2} \times 36.4$

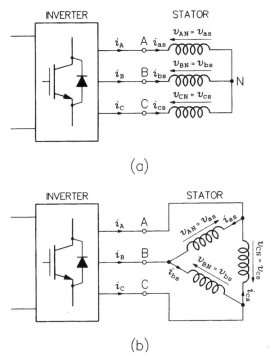

FIGURE 6.2   Stator currents and voltages: (a) wye-connected stator, (b) delta-connected stator.

$\cos(377t + 173.6°) = 51.5 \cos(377t + 173.6°)$ A; $i_{br} = 51.5 \cos(377t + 53.6°)$ A; and $i_{cr} = 51.5 \cos(377t - 66.4°)$ A.

At $t = 0$, individual currents are: $i_{as} = 50.0$ A, $i_{bs} = -46.6$ A, $i_{cs} = -3.4$ A, $i_{ar} = -51.2$ A, $i_{br} = 30.6$ A, and $i_{cs} = -20.6$ A. Eq. (6.5) yields $i_{ds} = 75.0$ A, $i_{qs} = -37.4$ A, $i_{dr} = -76.8$ A, and $i_{qr} = 8.7$ A. Thus, the space vectors of the stator and rotor currents are: $i_s = 75.0 - j37.4$ A $= 83.8 \angle -26.5°$ A and $i_r = -76.8 + j8.7$ A $= 77.3 \angle 173.6°$ A. Note the formal similarity between the phasors and space vectors: The magnitude of the vector is $1.5\sqrt{2}$ times greater than that of the rms phasor (or 1.5 times greater than that of the peak-value phasor), while the phase angle is the same for both quantities. ■

6.2 DYNAMIC EQUATIONS OF THE INDUCTION MOTOR

The dynamic T-model of the induction motor in the stator reference frame, with motor variables expressed in the vector form, is shown in Figure

6.3. Symbol p (not to be confused with the number of pole pairs, p_p) denotes the differentiation operator, d/dt, while L_{ls}, L_{lr}, and L_m are the stator and rotor leakage inductances and the magnetizing inductance, respectively ($L_{ls} = X_{ls}/\omega$, $L_{rs} = X_{rs}/\omega$, $L_m = X_m/\omega$). The sum of the stator leakage inductance and magnetizing inductance is called the *stator inductance* and denoted by L_s. Analogously, the *rotor inductance*, L_r, is defined as the sum of the rotor leakage inductance and magnetizing inductance. Thus, $L_s = L_{ls} + L_m$, and $L_r = L_{lr} + L_m$ ($L_s = X_s/\omega$, $L_r = X_r/\omega$).

The dynamic model allows derivation of the voltage-current equation of the induction motor. Using space vectors, the equation can be written as

$$\frac{di}{dt} = Av + Bi, \tag{6.7}$$

where

$$i = [i_{ds}\ i_{qs}\ i_{dr}\ i_{qr}]^T, \tag{6.8}$$

$$v = [v_{ds}\ v_{qs}\ v_{dr}\ v_{qr}]^T, \tag{6.9}$$

$$A = \frac{1}{L_\sigma^2}\begin{bmatrix} L_r & 0 & -L_m & 0 \\ 0 & L_r & 0 & -L_m \\ -L_m & 0 & L_s & 0 \\ 0 & -L_m & 0 & L_s \end{bmatrix}, \tag{6.10}$$

$$B = B(\omega_o) = \frac{1}{L_\sigma^2}\begin{bmatrix} -R_s L_r & \omega_o L_m^2 & R_r L_m & \omega_o L_r L_m \\ -\omega_o L_m^2 & -R_s L_r & -\omega_o L_r L_m & R_r L_m \\ R_s L_m & -\omega_o L_s L_m & -R_r L_s & -\omega_o L_s L_r \\ \omega_o L_s L_m & R_s L_m & \omega_o L_s L_r & -R_r L_s \end{bmatrix}, \tag{6.11}$$

$$L_\sigma^2 = L_s L_r - L_m^2. \tag{6.12}$$

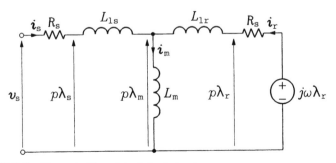

FIGURE 6.3   Dynamic T-model of the induction motor.

Symbols i_{dr} and i_{qr} in Eq. (6.9) denote components of the *rotor current vector*, i_r. In the squirrel-cage motor, the corresponding components, v_{dr} and v_{qr}, of the *rotor voltage vector*, v_r, are both zero because the rotor windings are shorted.

The stator and rotor fluxes are related to the stator and rotor current, as

$$\begin{bmatrix} \lambda_s \\ \lambda_r \end{bmatrix} = \begin{bmatrix} L_s & L_m \\ L_m & L_r \end{bmatrix} \begin{bmatrix} i_s \\ i_r \end{bmatrix}. \tag{6.13}$$

The stator flux can also be obtained from the stator voltage and current as

$$\frac{d\lambda_s}{dt} = v_s - R_s i_s \tag{6.14}$$

or

$$\lambda_s = \int_0^t (v_s - R_s i_s)dt + \lambda_s(0), \tag{6.15}$$

while the rotor flux in the squirrel-cage motor satisfies the equation

$$\frac{d\lambda_r}{dt} = j\omega_o\lambda_r - R_r i_r \tag{6.16}$$

Finally, the developed torque can be expressed in several forms, such as

$$T_M = \frac{2}{3}p_p Im\{i_s \lambda_s^*\} = \frac{2}{3}p_p(i_{qs}\lambda_{ds} - i_{ds}\lambda_{qs}), \tag{6.17}$$

$$T_M = \frac{2}{3}p_p\frac{L_m}{L_r}Im\{i_s \lambda_r^*\} = \frac{2}{3}p_p\frac{L_m}{L_r}(i_{qs}\lambda_{qr} - i_{ds}\lambda_{qr}), \tag{6.18}$$

or

$$T_M = \frac{2}{3}p_p L_m Im\{i_s i_r^*\} = \frac{2}{3}p_p L_m(i_{qs}i_{dr} - i_{ds}i_{qr}), \tag{6.19}$$

where the star denotes a conjugate vector.

The rather abstract term $Im(i_s\lambda_s^*)$ in Eq. (6.17) and the analogous terms in Eqs. (6.18) and (6.19) represent a vector product of the involved space vectors. For instance,

$$Im(i_s\lambda_s^*) = i_s\lambda_s\sin[\angle(i_s,\lambda_s)]. \tag{6.20}$$

Eq. (6.20) implies that the torque developed in an induction motor is proportional to the product of magnitudes of space vectors of two selected motor variables (two currents, two fluxes, or a current and a flux) and the sine of angle between these two vectors. It can be seen that all the torque equations are nonlinear, as each of them includes a difference of products of two motor variables. Eq. (6.7) is nonlinear, too, because of the variable ω_o appearing in matrix B.

EXAMPLE 6.2   In this example, the stator and rotor fluxes under rated operating conditions will first be calculated and followed by torque calculations using various formulas. All these quantities are constant in time, thus the instant $t = 0$ can be considered. For this instant, from Example 6.1, $i_s = 75.0 - j37.4$ A $= 83.8 \angle -26.5°$ A and $i_r = -76.8 + j8.7$ A $= 77.3 \angle 173.6°$ A. From Eq. (6.13), $\lambda_s = 0.04424 \times (75.0 - j37.4) + 0.041 \times (-76.8 + j8.7) = 0.032 - j.229$ Wb $= 1.229 \angle -88.5°$ Wb and $\lambda_r = 0.041 \times (75.0 - j37.4) + 0.0417 \times (-76.8 + j8.7) = -0.128 - j.171$ Wb $= 1.178 \angle -96.2°$ Wb. Thus, the rated rms value of the stator flux, Λ_s, is $1.229/(1.5\sqrt{2}) = 0.580$ Wb (a similar value was already employed in Example 5.1) and that, Λ_r, of the rotor flux is $1.178/(1.5\sqrt{2}) = 0.555$ Wb.

From Eq. (6.17), $T_M = 2/3 \times 3 \times \text{Im}\{83.8\angle -26.5° \times 1.229 \angle 88.5°\} = 2 \times \text{Im}\{103\angle 62°\} = 2 \times 103 \times \sin(62°) = 181.9$ Nm. Analogously, from Eq. (6.18), $T_M = 2/3 \times 0.041/0.0417 \times 3 \times \text{Im}\{83.8\angle -26.5° \times 1.178 \times \angle 96.2°\} = 182.1$ Nm, and from Eq. (6.19), $T_M = 2/3 \times 0.041 \times 3 \times \text{Im}\{83.8\angle -26.5° \times 77.3 \angle -173.6°\} = 182.5$ Nm. All three results are very close to the value of 183.1 Nm obtained in Example 5.1 (the differences are due only to round-up errors).   ∎

6.3 REVOLVING REFERENCE FRAME

In the steady state, space vectors of motor variables revolve in the stator reference frame with the angular velocity, ω, imposed by the supply source (inverter). It must be stressed that this velocity does not depend on the number of poles of stator, which indicates the somewhat abstract quality of the vectors (the speed of the actual stator MMF, a "real" space vector, equals ω/p_p). Under transient operating conditions, instantaneous speeds of the space vectors vary, and they are not necessarily the same for all vectors, but the vectors keep revolving nevertheless. Consequently, their d and q components are ac variables, which are less convenient to analyze and utilize in a control system than the dc signals commonly used in control theory. Therefore, in addition to the *static*, abc→dq and dq→abc, transformations, the *dynamic*, dq→DQ and DQ→dq, transformations from the stator reference frame to a revolving frame and vice versa are often employed. Usually, the revolving reference frame is so selected that it moves in synchronism with a selected space vector.

The revolving reference frame, DQ, rotating with the frequency ω_e (the subscript "e" comes from the commonly used term "excitation frame"), is shown in Figure 6.4 with the stator reference frame in the background. The stator voltage vector, $\mathbf{v}_s$, revolves in the stator frame with the angular velocity of ω, remaining stationary in the revolving frame if $\omega_e = \omega$. Consequently, the v_{DS} and v_{QS} components of that vector in the latter frame are dc signals, constant in the steady state and varying in transient states. Considering the same stator voltage vector, its dq→DQ transformation is given by

$$\begin{bmatrix} v_{DS} \\ v_{QS} \end{bmatrix} = \begin{bmatrix} \cos(\omega_e t) & \sin(\omega_e t) \\ -\sin(\omega_e t) & \cos(\omega_e t) \end{bmatrix} \begin{bmatrix} v_{ds} \\ v_{qs} \end{bmatrix} \tag{6.21}$$

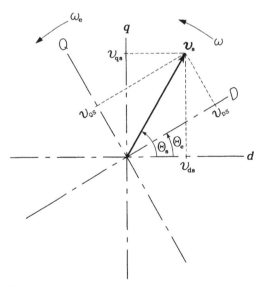

FIGURE 6.4   Space vector of stator voltage in the stationary and revolving reference frames.

and the inverse, DQ→dq, transformation by

$$\begin{bmatrix} v_{ds} \\ v_{qs} \end{bmatrix} = \begin{bmatrix} \cos(\omega_e t) & -\sin(\omega_e t) \\ \sin(\omega_e t) & \cos(\omega_e t) \end{bmatrix} \begin{bmatrix} v_{DS} \\ v_{QS} \end{bmatrix}. \qquad (6.22)$$

EXAMPLE 6.3 The stator current vector, i_s, from Example 6.1 is considered here to illustrate the ac-to-dc transformation of motor quantities realized by the use of revolving reference frame. At $t = 0$, $i_s(0) = 75.0 - j37.4$ A $= 83.8 \angle -26.5°$ A. Thus, recalling that in the steady state the space vectors rotate in the stator reference frame with the angular velocity ω_e equal ω, the time variations of the stator current vector can be expressed as $i_s(t) = 83.8 \exp[j(\omega t - 26.5°)]$ $= 83.8 \cos(\omega t - 26.5°) + j83.8 \sin(\omega t - 26.5°)$. Both components of the stator current vector are thus sinusoidal ac signals.

If the D axis of the reference frame is aligned with i_s, then, according to Eq. (6.21) (adapted to the stator current vector), $i_{DS} = \cos(\omega t) \times 83.8 \cos(\omega t - 26.5°) + \sin(\omega t) \times 83.8 \sin(\omega t - 26.5°)$ $= 83.8[\cos(\omega t)\cos(\omega t - 26.5°) + \sin(\omega t)\sin(\omega t - 26.5°)] = 83.8$ $\cos(\omega t - \omega t + 26.5°) = 83.8 \cos(26.5°) = 75.0$ A, and $i_{QS} = -\sin(\omega t) \times 83.8 \cos(\omega t - 26.5°) + \cos(\omega t) \times 83.8 \sin(\omega t - 26.5°)$ $= 83.8[\sin(\omega t - 26.5°)\cos(\omega t) - \cos(\omega t - 26.5°)\sin(\omega t)] = 83.8$ $\sin(\omega t - 26.5° - \omega t) = 83.8 \sin(-26.5°) = -37.4$ A.

It can be seen that $i_{DS} = i_{ds}(0)$ and $i_{QS} = i_{qs}(0)$. It would not be so if the revolving reference frame were aligned with another vector, but i_{DS} and i_{QS} would still be dc signals. ∎

To indicate the reference frame of a space vector, appropriate superscripts are used. For instance, the stator voltage vector in the stator reference frame can be expressed as

$$v_s^s = v_{ds} + jv_{qs} = v_s e^{j\Theta_s}, \qquad (6.23)$$

and the same vector in the revolving frame as

$$v_s^e = v_{DS} + jv_{QS} = v_s e^{j(\Theta_s - \Theta_e)}, \qquad (6.24)$$

where Θ_e denotes the angle between the frames. Angles Θ_s and Θ_e are given by

$$\Theta_s = \int_0^t \omega dt + \Theta_s(0) \qquad (6.25)$$

and

$$\Theta_e = \int_0^t \omega_e dt + \Theta_e(0). \tag{6.26}$$

Motor equations in a reference frame revolving with the angular velocity of ω_e can be obtained from those in the stator frame by replacing the differentiation operator p with $p + j\omega_e$. In particular,

$$\frac{d\boldsymbol{\lambda}_s^e}{dt} = \boldsymbol{v}_s^e - R_s\boldsymbol{i}_s^e - j\omega_e\boldsymbol{\lambda}_s^e \tag{6.27}$$

and

$$\frac{d\boldsymbol{\lambda}_r^e}{dt} = -R_r\boldsymbol{i}_r^e - j(\omega_e - \omega_o)\boldsymbol{\lambda}_r^e. \tag{6.28}$$

Equations that do not involve differentiation or integration, such as the torque equations, are the same in both frames.

6.4 SUMMARY

Three-phase variables in the induction motor can be represented by space vectors in the Cartesian coordinate set, dq, affixed to stator (stator reference frame). Space vectors of the stator voltage and current and magnetic fluxes (flux linkages) are commonly employed in the analysis and control of induction motor ASDs. The space vectors are obtained by an invertible, static, abc→dq, transformation of phase variables. The vector notation is used in dynamic equations of the motor.

Space vectors in the stator reference frame are revolving, so that their d and q components are ac signals. A dynamic, dq→DQ, transformation allows conversion of those signals to the dc form. The dq→DQ transformation introduces a revolving frame of reference, in which, in the steady state, space vectors appear as stationary. The revolving frame is usually synchronized with a space vector of certain motor variable. Dynamic equations of the induction motor can easily be adapted to a revolving reference frame by substituting $p + j\omega_e$ for p.

7

FIELD ORIENTATION

This chapter begins with a review of conditions of production and control of torque in the dc motor. Principles of the field orientation in the induction motor are introduced, and the direct and indirect field orientation schemes utilizing the rotor flux vector are presented. Stator and air-gap flux orientation systems are described, and we finish with an explanation of the control of stator current in field-oriented motors fed from current source inverters.

7.1 TORQUE PRODUCTION AND CONTROL IN THE DC MOTOR

The concept of field orientation, proposed by Hasse in 1969 and Blaschke in 1972, constitutes, arguably, the most important paradigm in the theory and practice of control of induction motors. In essence, the objective of field orientation is to make the induction motor emulate the separately excited dc machine as a source of adjustable torque. Therefore, we will first review fundamentals of torque production and control in the dc motor.

119

A simplified representation of the dc motor is shown in Figure 7.1. The pair of magnetic poles, N and S, represent the magnetic circuit of stator, that is, the field part of the machine. Therefore, the space vector, λ_f, of flux (flux linkage) generated by the field winding, is stationary and aligned with the d axis of the stator. Thanks to the action of commutator (not shown) and properly positioned brushes, the distribution of armature current in the rotor winding is such that the space vector, i_a, of this current is always aligned with the q axis, even though the rotor is revolving.

Practical dc motors are equipped with auxiliary windings designated to neutralize the so-called *armature reaction*, that is, weakening of the main magnetic field by the MMF produced by the armature current. Then, the developed torque, T_M, is proportional to the vector product of i_a and λ_f, that is, to the sine of angle between these vectors. As seen in Figure 7.1, this is always a right angle, which ensures the highest *torque-per-ampere ratio*. Thus, the torque is produced under optimal conditions, the minimum possible armature current causing minimum losses in the motor and supply system.

In the separately excited dc machine, the field current, i_f, producing λ_f and the armature current, i_a, flow in separate windings. Therefore, they can be controlled independently. Usually, particularly under high-load operating conditions, the flux is kept constant at the rated level within the speed range of zero to the rated value. Field weakening, that is, flux reduction in inverse proportion to the speed, is used at speeds higher than rated. With low loads, the motor efficiency can be improved by reducing the flux to such a trade-off value that the resultant decrease in core losses offsets the simultaneous increase in copper losses. In drives that most of

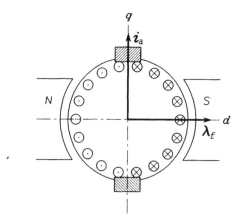

FIGURE 7.1 Simplified representation of the dc motor.

the time run well under the full load, such efficiency optimization schemes can bring significant energy savings.

Presented considerations explain why the separately excited dc motor were for many decades the favorite actuator in motion control systems. The dc motor not only generates the torque under the optimal condition of orthogonality of the flux and current vectors, but it also allows fully independent ("decoupled") control of the torque and the magnetic field. The torque equation, to be used as a reference for the field-oriented induction motor, can be written as

$$T_M = k_T \lambda_f i_a, \tag{7.1}$$

where k_T is a constant dependent on the construction and size of the motor.

7.2 PRINCIPLES OF FIELD ORIENTATION

All three torque equations, (6.17) through (6.19), of the induction motor in the stator reference frame include the difference-of-products terms. Notice that if, for example, in Eq. (6.18), λ_{qr} were made to equal zero, the resultant formula would be similar to that, (7.1), for the dc motor. Unfortunately, this is not possible, because λ_{qr} constitutes the quadrature component of the revolving vector, $\boldsymbol{\lambda}_r$, of rotor flux. Thus, $\lambda_{qr} = 0$ is possible only if $\boldsymbol{\lambda}_r = 0$, which is absurd.

However, if torque equations in a revolving frame are considered, the manipulation described above becomes feasible. If

$$T_M = \frac{2}{3} p_P \frac{L_m}{L_r} (i_{QS} \lambda_{DR} - i_{DS} \lambda_{QR}) \tag{7.2}$$

and $\lambda_{DR} = 0$, then

$$T_M = k_T \lambda_{DR} i_{QS}, \tag{7.3}$$

where $k_T = 2p_p L_m/(3L_r)$, and the induction motor, as desired, emulates the dc machine. The described condition is realized by aligning the D axis of the revolving reference frame with the rotor flux vector, $\boldsymbol{\lambda}_r$, as illustrated in Figure 7.2. Similar results can be obtained aligning the D axis with another flux vector, that is, the stator or air-gap flux vector. For instance, the stator field orientation, according to Eq. (6.17), yields

$$T_M = k'_T \lambda_{DS} i_{QS} \tag{7.4}$$

where $k'_T = 2p_p/3$.

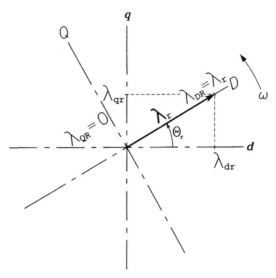

FIGURE 7.2 Alignment of the revolving reference frame with the rotor flux vector.

Principles of field orientation along a selected flux vector, λ_f (λ_r, λ_s, or λ_m), can be summarized as follows.

1. Given the reference values, T_M^* and λ_f^*, of the developed torque and selected flux, find the corresponding reference components, i_{DS}^* and i_{QS}^*, of the stator current vector in the revolving reference frame.

2. Determine the angular position, Θ_f, of the flux vector in question, to be used in the DQ→dq conversion from i_{DS}^* to i_{ds}^* and from i_{QS}^* to i_{qs}^*.

3. Given the reference components, i_{ds}^* and i_{qs}^*, of the stator vector in the stator reference frame, use the dq→abc transformation to obtain reference stator currents, i_{as}^*, i_{bs}^*, and i_{cs}^*, for a current-controlled inverter feeding the motor.

Based on the dynamic equations of the induction motor, a block diagram of the field-oriented motor in a revolving reference frame aligned with the rotor flux vector is shown in Figure 7.3. According to the theory of linear dynamic systems, the integrator block with negative feedback can be replaced with a first-order block, as shown in Figure 7.4. It can be seen that the torque in a field-oriented motor reacts instantly to changes in the i_{QS} component of the stator current, while the reaction of rotor flux to changes in the other component, i_{DS}, is inertial.

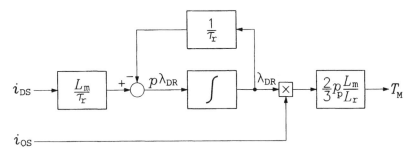

FIGURE 7.3 Block diagram of the field-oriented motor in a revolving reference frame aligned with the rotor flux vector.

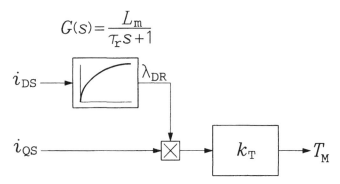

FIGURE 7.4 Reduced block diagram of the field-oriented motor in a revolving reference frame aligned with the rotor flux vector.

Because $\omega_e = \omega$ and $\omega_e - \omega_o = \omega_r$, then the real part of Eq. (6.28) may be written as

$$i_{DR} = \frac{1}{R_r}\left(\omega_r \lambda_{QR} - \frac{d\lambda_{DR}}{dt}\right). \tag{7.5}$$

Under the field orientation condition, $\lambda_{QR} = 0$ and, with $\lambda_{DR} = const$, $d\lambda_{DR}/dt = 0$, too. Hence, $i_{DR} = 0$ and $i_r^e = i_{QR}$, which, because $\lambda_r^e = \lambda_{DR}$, indicates orthogonality of the rotor current and flux vectors. This is the condition of optimal torque production, that is, the maximum torque per ampere ratio, typical for the dc motor. Thus, the field orientation makes operating characteristics of the induction motor similar to those of that machine.

7.3 DIRECT FIELD ORIENTATION

Knowledge of the instantaneous position (angle) of the flux vector, with which the revolving reference frame is aligned, constitutes the necessary requirement for proper field orientation. Usually, the magnitude of the flux vector in question is identified as well, for comparison with the reference value in a closed-loop control scheme. Identification of the flux vector can be based on direct measurements or estimation from other measured variables. Such an approach is specific for schemes with the so-called *direct field orientation* (DFO), which will be explained for the rotor flux vector, $\boldsymbol{\lambda}_r$, as the orienting vector.

Only the air-gap flux can be measured directly. A simple scheme for estimation of the rotor flux vector, based on measurements of the airgap flux and stator currents, is depicted in Figure 7.5. Two Hall sensors of magnetic field are placed in the motor gap, measuring the direct and quadrature components, λ_{dm} and λ_{qm}, of the air-gap flux vector, $\boldsymbol{\lambda}_m$. Stator currents are measured too. The rotor flux vector, in the rectangular or polar form, is calculated as

$$\boldsymbol{\lambda}_r = \frac{L_r}{L_m}\boldsymbol{\lambda}_m - L_{ls}\boldsymbol{i}_s. \tag{7.6}$$

As an alternative to the fragile Hall sensors, flux sensing coils or taps on the stator winding can be installed in the motor. Voltages induced in the coils or the winding are integrated to provide λ_{dm} and λ_{qm}.

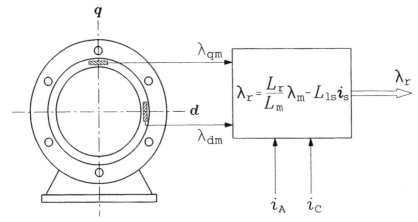

FIGURE 7.5   Estimation of the rotor flux vector based on direct measurements of the air-gap flux.

Sensors of the air-gap flux are inconvenient, and they spoil the ruggedness of the induction motor. Therefore, in practice, the rotor flux vector (or another flux vector used for the field orientation) is usually computed from the stator voltage and current. In particular, the stator flux vector, λ_s, can be estimated using Eq. (6.15) which, in turn, allows calculation of the air-gap flux vector, λ_m, as

$$\lambda_m = \lambda_s - L_{ls}i_s \tag{7.7}$$

and estimation of the rotor flux vector, λ_r, from Eq. (7.6).

For best performance, the torque and flux in induction motors with direct field orientation are closed-loop controlled. The torque, which is difficult to measure directly, can be calculated using an appropriate equation, such as (6.18). A block diagram of the ASD with direct rotor flux orientation using air-gap flux sensors is shown in Figure 7.6. Proportional-integral (PI) controllers used in loops of the flux and torque control generate reference components, i_{DS}^* and i_{QS}^*, of the stator current vector in the revolving reference frame. The DQ→dq dynamic transformation block converts the i_{DS}^* and i_{QS}^* dc signals into i_{ds}^* and i_{qs}^* ac signals representing reference components of the stator current vector in the stator reference frame. Operation of the dynamic transformation block is synchronized by the angle signal, Θ_r, from the flux calculator. The i_{ds}^* and i_{qs}^* signals are applied to the dq→abc static transformation block to produce reference currents, i_{as}^*, i_{bs}^* and i_{cs}^*, for individual phases of the current-controlled inverter.

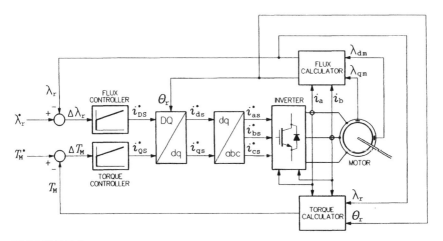

FIGURE 7.6   Block diagram of the ASD with direct rotor flux orientation.

In the block diagram in Figure 7.6 and throughout this whole book, the generic term "calculator" is used to represent an open-loop *estimator* or closed-loop *observer*. Chapter 10 contains more information on the estimators and observers of motor quantities.

EXAMPLE 7.1   At a certain instant, the angle, Θ_r, of the rotor flux vector in the drive in Figure 7.6 is 130°. The output signals, i_{DS}^* and i_{QS}^*, from the flux and torque controllers are 20 A and −30 A, respectively. Find the required (reference) values of individual phase currents in the stator.

Based on the DQ→dq transformation (6.22), $i_{ds}^* = 20 \cos(130°) + 30 \sin(130°) = 10.1$ A, and $i_{qs}^* = 20 \sin(130°) + 30 \cos(130°) = -4.0$ A. Finally, the dq→abc transformation (6.6) yields $i_{as}^* = 2/3 \times 10.1 = 6.7$ A, $i_{bs}^* = -1/3 \times 10.1 - 1/\sqrt{3} \times 4.0 = -5.7$ A, and $i_{cs}^* = -1/3 \times 10.1 + 1/\sqrt{3} \times 4.0 = -1.0$ A (the last result could also be obtained from the Kirchhoff current law: $i_{cs}^* = -i_{as}^* - i_{cs}^* = -6.7 + 5.7 = -1.0$ A).   ∎

7.4 INDIRECT FIELD ORIENTATION

In an alternative approach to direct flux orientation, the *indirect field orientation* (IFO), the angular position, Θ_r, of the rotor flux vector is determined indirectly as

$$\Theta_r = \int_0^t \omega_r^* dt + p_p \Theta_M, \tag{7.8}$$

where ω_r^* denotes the rotor frequency required for field orientation and Θ_M is the angular displacement of the rotor, measured by a shaft position sensor, typically a digital encoder. The required rotor frequency can be computed directly from motor equations under the field orientation condition. With $\lambda_r^e = \lambda_{DR}$,

$$i_r^e = \frac{1}{L_r}(\lambda_{DR} - L_m i_s^e), \tag{7.9}$$

which, when substituted in Eq. (6.28), yields

$$\lambda_{DR}[1 + \tau_r(p + j\omega_r)] = L_m i_s^e, \tag{7.10}$$

where τ_r denotes the rotor time constant, L_r/R_r. Splitting Eq. (7.10) into the real and imaginary parts gives

$$\lambda_{DR}(1 + p\tau_r) = L_m i_{DS} \tag{7.11}$$

and

$$\omega_r \tau_r \lambda_{DR} = L_m i_{QS}. \qquad (7.12)$$

Replacing ω_r with ω_r^*, λ_{DR} with λ_r^*, and i_{QS} with i_{QS}^* in the last equation, and solving for ω_r^*, yields

$$\omega_r^* = \frac{L_m}{\tau_r} \frac{i_{QS}^*}{\lambda_r^*}. \qquad (7.13)$$

Note that Eq. (7.13) is a time-domain equivalent of the phasor-domain Eq. (5.27) for the scalar torque control. Indeed, from Eq. (7.11), in the steady state of the motor ($p = 0$),

$$\lambda_r^* = \lambda_{DR}^* = L_m i_{DS}^*, \qquad (7.14)$$

which, when substituted in Eq. (7.13), gives

$$\omega_r^* = \frac{1}{\tau_r} \frac{i_{QS}^*}{i_{DS}^*}. \qquad (7.15)$$

Variables i_{DS}^* and i_{QS}^* represent the required flux-producing and torque-producing components of the stator current vector, i_s^*, in the same way that I_Φ^* and I_M^* in Eq. (5.27) are the respective components of the stator current *phasor*, I_s^*.

The reference current i_{DS}^* corresponding to a given reference flux, λ_r^*, can be found from Eq. (7.11) as

$$i_{DS}^* = \frac{\tau_r p + 1}{L_m} \lambda_r^* = \frac{1}{L_m} \left(\tau_r \frac{d\lambda_r^*}{dt} + \lambda_r^* \right), \qquad (7.16)$$

while the other reference current, i_{QS}^*, for a given reference torque, T_M^*, can be obtained from the torque equation (7.3) of a field-oriented motor as

$$i_{QS}^* = \frac{1}{k_T} \frac{T_M^*}{\lambda_r^*}. \qquad (7.17)$$

A drive system with indirect rotor flux orientation is shown in Figure 7.7. In accordance with Eq. (7.8), the angle, Θ_r, of the rotor flux vector used in the DQ→dq transformation is determined as

$$\Theta_r = \Theta^* + \Theta_o, \qquad (7.18)$$

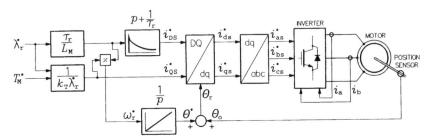

FIGURE 7.7   Block diagram of the ASD with indirect rotor flux orientation.

where Θ^* denotes the time integral of the reference rotor frequency, ω_r^*, and $\Theta_o = p_p \Theta_M$ is the angular displacement of the rotor in an equivalent two-pole motor.

EXAMPLE 7.2   The example motor operates in the indirect field orientation scheme, with the flux and torque commands equal to the respective rated values, that is, $\lambda_r^* = 1.178$ Wb (see Example 6.2) and $T_M^* = 183$ Nm. At the instant $t = 1$ s since starting the motor, the rotor has made 8 revolutions. Determine the reference output currents, i_{as}^*, i_{bs}^*, and i_{cs}^*, of the inverter as calculated by the digital control system.

The torque coefficient, $k_T = 2p_p L_m/(3L_r)$, is $2 \times 3 \times 0.041/(3 \times 0.0417) = 1.97$. Because $\lambda_r^* = $ const, current i_{DS}^* can be found from Eq. (7.14) as $i_{DS}^* = \lambda_r^*/L_m = 1.178/0.041 = 28.7$ A, while current i_{QS}^*, from Eq. (7.17), is $T_M^*/(k_T \lambda_r^*) = 183/(1.97 \times 1.178) = 78.9$ A. The rotor time constant, τ_r, is given by $\tau_r = L_r/R_r = 0.0417/0.156 = 0.267$ s, and the reference rotor frequency, ω_r^*, can be calculated from Eq. (7.15) as $i_{QS}^*/(\tau_r i_{DS}^*) = 78.9/(0.267 \times 28.7) = 10.3$ rad/s. This result can be verified by finding the rated slip, s_{rat}, at the rated supply frequency of 60 Hz (377 rad/s) as $10.3/377 = 0.027$, a value already listed in Table 2.1.

The integral, Θ^*, of ω_r^* at $t = 1$ s is $\Theta^* = \omega_r^* t = 10.3 \times 1 = 10.3$ rad, while the rotor displacement, Θ_o, of the equivalent 2-pole motor equals p_p of the displacement of the actual motor, that is, $\Theta_o = 3 \times 8 \times 2\pi = 150.8$ rad. Thus, angle Θ_r, for the DQ→dq transformation is $\Theta_r = \Theta^* + \Theta_o = 10.3 + 150.8 = 161.1$ rad, which is equivalent to $230.4°$. Now, from Eq. (6.22),

$$\begin{bmatrix} i_{ds}^* \\ i_{qs}^* \end{bmatrix} = \begin{bmatrix} \cos(230.4°) & -\sin(230.4°) \\ \sin(230.4°) & \cos(230.4°) \end{bmatrix} \begin{bmatrix} 28.7 \\ 78.9 \end{bmatrix} = \begin{bmatrix} 42.5\ A \\ -72.4\ A \end{bmatrix},$$

and, from Eq. (6.6),

$$
\begin{bmatrix} i^*_{as} \\ i^*_{bs} \\ i^*_{cs} \end{bmatrix} = \begin{bmatrix} \dfrac{2}{3} & 0 \\ -\dfrac{1}{3} & \dfrac{1}{\sqrt{3}} \\ -\dfrac{1}{3} & -\dfrac{1}{\sqrt{3}} \end{bmatrix} \begin{bmatrix} 42.5 \\ -72.4 \end{bmatrix} = \begin{bmatrix} 28.3\ A \\ -55.9\ A \\ 27.6\ A \end{bmatrix}.
$$

Note that the dq→DQ transformation does not affect the magnitude of a vector. Indeed, both i^s_s and i^e_s have the same magnitude of $[28.7^2 + (-78.9)^2]^{1/2} = [42.5^2 + (-72.4)^2]^{1/2} = 84.0$ A. ∎

7.5 STATOR AND AIR-GAP FLUX ORIENTATION

As subsequently demonstrated, stator and air-gap flux orientation systems require somewhat more complicated control algorithms than that for the rotor flux orientation described in the preceding sections. On the other hand, accurate estimation of the stator flux vector is easier than that of the rotor flux vector, while the air-gap flux vector, as mentioned in Section 7.3, can be measured directly.

Eq. (6.17) yields the condition for stator flux orientation as $\lambda_{QS} = 0$. Then, the developed torque is given by Eq. (7.4). Eq. (6.13) allows expressing the rotor current vector as

$$
i_r = \frac{1}{L_m}(\lambda_s - L_s i_s)
\tag{7.19}
$$

and the rotor flux vector as

$$
\lambda_r = \frac{L_r}{L_m}(\lambda_s - \sigma L_s i_s),
\tag{7.20}
$$

where

$$
\sigma = 1 - \frac{L_m^2}{L_s L_r}
\tag{7.21}
$$

denotes the so-called *total leakage factor* of the motor. When $\lambda_{QS} = 0$, then

$$
i^e_r = \frac{1}{L_m}[\lambda_{DS} - L_s(i_{DS} + ji_{QS})]
\tag{7.22}
$$

and

$$\lambda_r^e = \frac{L_r}{L_m}[\lambda_{DS} - \sigma L_s(i_{DS} + ji_{QS})]. \tag{7.23}$$

Substituting Eqs. (7.22) and (7.23) in Eq. (6.28), with $\omega_e = \omega$, that is, $\omega_e - \omega_o = \omega_r$, and solving for λ_{DS}, gives, after some rearrangements,

$$\lambda_{DS} = L_s \frac{1 + \sigma \tau_r(p + j\omega_r)}{1 + \tau_r(p + j\omega_r)}(i_{DS} + ji_{QS}). \tag{7.24}$$

Clearly, in a field-oriented motor, the λ_{DS} component of the stator flux vector may not have the imaginary part implied by Eq. (7.24). Also, the magnitude, $\lambda_s = \lambda_{DS}$, of this vector should be independent of the torque-producing component, i_{QS}, of the stator current vector. Therefore, the reference signal, i_{DS}^*, of the flux-producing current must be made dynamically dependent on the reference signal, i_{QS}^*, of the torque-producing current in such a way that i_{QS}^* has no effect on λ_{DS} and that λ_{DS} is a real number.

Separation of Eq. (7.24) into the real and imaginary parts yields

$$\lambda_{DS} = L_s\left(\frac{1 + \sigma \tau_r p}{1 + \tau_r p}i_{DS} - \frac{\sigma \tau_r \omega_r}{1 + \tau_r p}i_{QS}\right) \tag{7.25}$$

and

$$\lambda_{QS} = L_s\left(\sigma i_{DS} + \frac{1 + \sigma \tau_r p}{\tau_r \omega_r}i_{QS}\right). \tag{7.26}$$

Solving Eq. (7.25) for i_{DS} and Eq. (7.26) for ω_r gives the reference values of these two variables as

$$i_{DS}^* = \frac{\left(p + \dfrac{1}{\tau_r}\right)\dfrac{\lambda_s^*}{\sigma L_s} + \omega_r^* i_{QS}^*}{p + \dfrac{1}{\sigma \tau_r}} \tag{7.27}$$

and

$$\omega_r^* = \frac{p + \dfrac{1}{\sigma \tau_r}}{\dfrac{\lambda_s^*}{\sigma L_s} - i_{DS}^*}i_{QS}^*. \tag{7.28}$$

The system for dynamic decoupling of the flux-producing and torque-producing currents is shown in Figure 7.8.

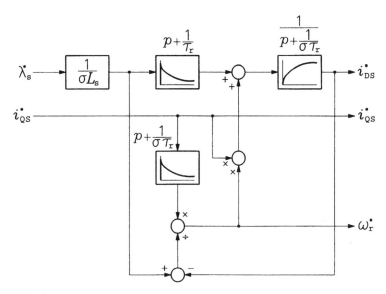

FIGURE 7.8 System for dynamic decoupling of the flux-producing and torque-producing currents.

Clearly, in practical systems, the reference torque and flux values must be limited to avoid current overloads. Aside from that, in the classic rotor flux orientation schemes described in Sections 7.3 and 7.4, there are no additional limits on the torque and flux commands. However, such limits do exist in drives with the stator flux orientation and air-gap flux orientation. The steady-state value, ω_r^*, of the rotor frequency command can be determined from Eqs. (7.27) and (7.28) by setting operator p to zero and eliminating i_{DS}^*. As a result, the following quadratic equation is obtained:

$$\omega_r^{*2} - \frac{(1 - \sigma)\lambda_s^*}{\sigma^2 L_s \tau_r i_{QS}^*}\omega_r + \frac{1}{(\sigma\tau_r)^2} = 0. \tag{7.29}$$

For the solution of Eq. (7.29) to be a real number, the discriminant may not be negative, which requires that

$$|i_{QS}^*| \leq \frac{\lambda_s^*}{2L_s}\left(\frac{1}{\sigma} - 1\right). \tag{7.30}$$

Because, from Eq. (7.4), $i_{QS}^* = T^*/(k_T'\lambda_s^*)$, then condition (7.30) can be rearranged to

$$\frac{|T_M^*|}{\lambda_s^{*2}} \leq \frac{k_T'}{2L_s}\left(\frac{1}{\sigma} - 1\right). \tag{7.31}$$

It can be seen that it is the total leakage factor, σ, that imposes the "torque-per-flux-squared" limitation in the stator flux orientation schemes.

EXAMPLE 7.3   Assuming the stator flux at the rated level of 1.229 Wb (see Example 6.2), what is the maximum allowable reference torque for the example motor in the stator field orientation scheme?

From Eq. (7.20), the total leakage factor, σ, is $1 - 0.041^2/(0.0424 \times 0.0417) = 0.049$, while $k_T' = 2 \times 3/3 = 2$ Nm/A.Wb. Thus, from condition (7.31), $|T_M'| < 2(1/0.049 - 1) \times 1.229^2/(2 \times 0.0424) = 691$ Nm. This limitation corresponds to 3.8 times the rated torque. ∎

The decoupling system in Figure 7.8 can be augmented to a reference current system shown in Figure 7.9. The latter system forms an important part of the direct field orientation scheme shown in Figure 7.10. The stator flux vector is calculated from Eq. (6.15). Alternatively, it can be determined from information about the air-gap flux and stator current, as

$$\boldsymbol{\lambda}_s = \boldsymbol{\lambda}_m + L_{ls}\boldsymbol{i}_s. \tag{7.32}$$

Knowledge of the stator flux vector allows calculation of the developed torque from Eq. (6.17). The easy identification of the stator flux vector makes an indirect stator flux orientation scheme superfluous.

In a similar way, the D axis of the revolving reference frame for transformation of signals i_{DS}^* and i_{QS}^* into signals i_{as}^*, i_{bs}^*, and i_{cs}^* can be

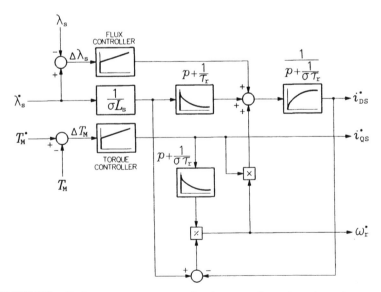

FIGURE 7.9   Reference current system for the stator flux orientation scheme.

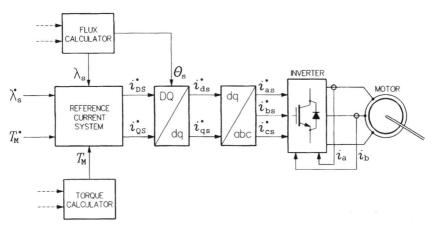

FIGURE 7.10   Block diagram of the ASD with stator flux orientation.

aligned with the air-gap flux vector, λ_m. The air-gap flux orientation scheme is similar to that of the stator flux orientation, with λ_s^* replaced with λ_m^*, λ_s with λ_m, L_s with L_m, and σ with σ_r, the latter symbol denoting the *rotor leakage factor*, defined as

$$\sigma_r = \frac{L_{lr}}{L_r}. \tag{7.33}$$

The air-gap flux vector is either measured directly or estimated from Eq. (7.7). The developed torque in an air-gap flux oriented motor is given by

$$T_M = k_T' i_{QS} \lambda_{DM} \tag{7.34}$$

with the same torque coefficient, $k_T' = 2p_p/3$, as that in Eq. (7.4). It can be demonstrated that the "torque-per-flux-squared" limitation is less restrictive than that, (7.31), for a motor with stator flux orientation.

EXAMPLE 7.4   Repeat Example 7.3 for the example motor with air-gap flux orientation.

Based on results of Example 6.2, the rated air-gap flux vector at $t = 0$ can be calculated from Eq. (7.35) as $\lambda_m = (0.032 - 0.00139 \times 75) + j(-1.229 + 0.00139 \times 37.4) = -0.072 - j1.179 = 1.179\angle -93.5°$ Wb. From Eq. (7.33), the rotor leakage factor, σ_r, is 0.00139/ 0.0417 = 0.033, while $k_T' = 2 \times 3/3 = 2$ Nm/A.Wb. Thus, from condition (7.31), $|T_M'| < 2(1/0.033 - 1) \times 1.179^2/(2 \times 0.041) = 993$ Nm. This value corresponds to 5.4 times the rated torque. Recall that the same ratio in the stator flux oriented drive in Example 7.4 was 3.8. ■

In today's ASDs, control algorithms are implemented in digital systems, mostly microcontrollers and digital signal processors (DSPs). Changes from one type of field orientation to another are easy, as they are realized in the software only.

7.6 DRIVES WITH CURRENT SOURCE INVERTERS

Although most practical ac ASDs employ PWM voltage source inverters, current source inverters are still used in high-power vector controlled drives. Typically, they are based on the relatively slow GTOs, for which the square-wave operation mode is very appropriate. Current source inverters are fed from controlled rectifiers allowing bidirectional power flow. The current source for the inverter (the rectifier with a current feedback and the dc-link reactor) automatically limits the short-circuit currents, increasing reliability of the system.

Current control in current source inverters differs from that in voltage source inverters. A given reference vector, $i_s^* = I_i^* \angle \Theta_s^*$, of stator current is realized by separately adjusting the dc-link current, i_i, in the supplying rectifier and selecting such a state of the inverter that results in a position, Θ_s, of the current vector closest to Θ_s^*. A simplified block diagram of control scheme of the current source inverter is shown in Figure 7.11. The reference stator current signals, i_{DS}^* and i_{QS}^*, from the field-orientation

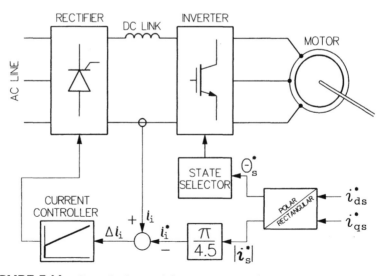

FIGURE 7.11   Control scheme of the current source inverter.

part of the control system of the ASD are converted to reference signals i_i^* and Θ_s^*. The dc-link current is controlled in the rectifier using a closed control loop with a PI controller, while the reference phase signals enforces an appropriate state of the inverter (see Figure 4.33).

Practical systems with current source inverters are more complicated, due to the problems of sluggish response to the magnitude command and the influence of stator EMF, $e_s = |v_s - R_s i_s|$, on the dc-link current, i_i. Lead-type compensators and EMF estimators are used for performance improvement.

7.7 SUMMARY

Field orientation, consisting in the alignment of a revolving reference frame with a space vector of selected flux, allows the induction motor to emulate the separately excited dc machine. In this machine, the magnetic field and developed torque can be controlled independently. In addition, the torque is produced under the optimal condition of orthogonality of the flux and current vectors, resulting in the maximum possible torque-per-ampere ratio.

In ASDs with direct field orientation along the rotor flux vector, λ_r, this vector is determined from direct measurements or estimations of the air-gap flux. The indirect field orientation is based on calculation of the angular position, Θ_r, of λ_r as a sum of an integral, Θ^*, of the rotor frequency, ω_r^*, required for the field orientation and the rotor angular displacement, Θ_o, of the equivalent two-pole motor.

Stator flux orientation has the advantage of employing the easily determinable stator flux vector. Disadvantages include the necessity of a decoupling system making the flux independent of the torque-producing current, as well as the "torque-per-flux-squared" limiting condition on the reference values of torque and stator flux. The field-orientation scheme using the air-gap flux vector is similar to that with the stator flux vector.

Voltage source inverters with closed-loop current control are employed in most field-oriented drives. In ASDs with current source inverters, the required space vector of the stator current is generated by simultaneous adjustment of the dc-link current and inverter state selection.

8

DIRECT TORQUE AND FLUX CONTROL

Control methods for high-performance ASDs with induction motors by direct selection of consecutive states of the inverter are presented in this chapter. The Direct Torque Control (DTC) and Direct Self-Control (DSC) techniques are explained, and we describe an enhanced version of the DTC scheme, employing the space-vector pulse width modulation in the steady state of the drive.

8.1 INDUCTION MOTOR CONTROL BY SELECTION OF INVERTER STATES

As shown in Chapter 7, induction motors in field-orientation ASDs are current controlled, that is, the control system produces reference values of currents in individual phases of the stator. Various current control techniques can be employed in the inverter supplying the motor, all of them based on the feedback from current sensors. Operation of the current control scheme results in an appropriate sequence of inverter states, so that the actual currents follow the reference waveforms.

Two ingenious alternative approaches to control of induction motors in high-performance ASDs make use of specific properties of these motors for direct selection of consecutive states of the inverter. These two methods of direct torque and flux control, known as the *Direct Torque Control* (*DTC*) and *Direct Self-Control* (*DSC*), are presented in the subsequent sections.

As already mentioned in Chapter 6, the torque developed in an induction motor can be expressed in many ways. One such expression is

$$T_M = \frac{2}{3}p_P\frac{L_m}{L_\sigma^2}Im(\lambda_s\lambda_r^*) = \frac{2}{3}p_P\frac{L_m}{L_\sigma^2}\lambda_s\lambda_r\sin(\Theta_{sr}), \tag{8.1}$$

where Θ_{sr} denotes the angle between space vectors, λ_s and λ_r, of stator and rotor flux, subsequently called a *torque angle*. Thus, the torque can be controlled by adjusting this angle. On the other hand, the magnitude, λ_s, of stator flux, a measure of intensity of magnetic field in the motor, is directly dependent on the stator voltage according to Eq. (6.15). To explain how the same voltage can also be employed to control Θ_{sr}, a simple qualitative analysis of the equivalent circuit of induction motor, shown in Figure 6.3, can be used.

From the equivalent circuit, we see that the derivative of stator flux reacts instantly to changes in the stator voltage, the respective two space vectors, v_s and $p\lambda_s$, being separated in the circuit by the stator resistance, R_s, only. However, the vector of derivative of the rotor flux, $p\lambda_r$, is separated from that of stator flux, $p\lambda_s$, by the stator and rotor leakage inductances, L_{ls} and L_{lr}. Therefore, reaction of the rotor flux vector to the stator voltage is somewhat sluggish in comparison with that of the stator flux vector. Also, thanks to the low-pass filtering action of the leakage inductances, rotor flux waveforms are smoother than these of stator flux.

The impact of stator voltage on the stator flux is illustrated in Figure 8.1. At a certain instant, t, the inverter feeding the motor switches to State 4, generating vector v_4 of stator voltage (see Figure 4.23). The initial vectors of stator and rotor flux are denoted by $\lambda_s(t)$ and λ_r, respectively. After a time interval of Δt, the new stator flux vector, $\lambda_s(t + \Delta t)$, differs from $\lambda_s(t)$ in both the magnitude and position while, assuming a sufficiently short Δt, changes in the rotor flux vector have been negligible. The stator flux has increased and the torque angle, Θ_{sr}, has been reduced by $\Delta\Theta_{sr}$. Clearly, if another vector of the stator voltage were applied, the changes of the stator flux vector would be different. Directions of change of the stator flux vector, λ_s, associated with the individual six nonzero

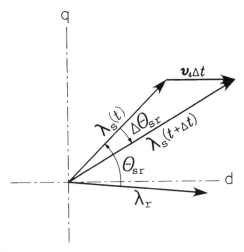

FIGURE 8.1   Illustration of the impact of stator voltage on the stator flux.

vectors, v_1 through v_6, of the inverter output voltage are shown in Figure 8.2, which also depicts the circular reference trajectory of λ_s. Thus, appropriate selection of inverter states allows adjustments of both the strength of magnetic field in the motor and the developed torque.

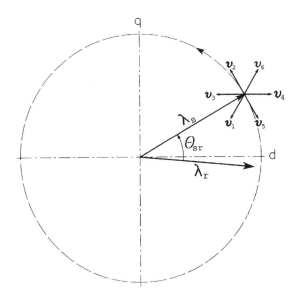

FIGURE 8.2   Illustration of the principles of control of stator flux and developed torque by inverter state selection.

8.2 DIRECT TORQUE CONTROL

The basic premises and principles of the Direct Torque Control (DTC) method, proposed by Takahashi and Noguchi in 1986, can be formulated as follows:

- Stator flux is a time integral of the stator EMF. Therefore, its magnitude strongly depends on the stator voltage.
- Developed torque is proportional to the sine of angle between the stator and rotor flux vectors.
- Reaction of rotor flux to changes in stator voltage is slower than that of the stator flux.

Consequently, both the magnitude of stator flux and the developed torque can be directly controlled by proper selection of space vectors of stator voltage, that is, selection of consecutive inverter states. Specifically:

- Nonzero voltage vectors whose misalignment with the stator flux vector does not exceed $\pm 90°$ cause the flux to increase.
- Nonzero voltage vectors whose misalignment with the stator flux vector exceeds $\pm 90°$ cause the flux to decrease.
- Zero states, 0 and 7, (of reasonably short duration) practically do not affect the vector of stator flux which, consequently, stops moving.
- The developed torque can be controlled by selecting such inverter states that the stator flux vector is accelerated, stopped, or decelerated.

For explanation of details of the DTC method, it is convenient to rename the nonzero voltage vectors of the inverter, as shown in Figure 8.3. The Roman numeral subscripts represent the progression of inverter states in the square-wave operation mode (see Figure 4.21), that is, $v_I = v_4$, $v_{II} = v_6$, $v_{III} = v_2$, $v_{IV} = v_3$, $v_V = v_1$, and $v_{VI} = v_5$. The Kth ($K = I$, II,..., VI) voltage vector is given by

$$v_K = V_i e^{j\Theta_{v,K}}, \tag{8.2}$$

where V_i denotes the dc input voltage of the inverter and

$$\Theta_{v,K} = (K - 1)\frac{\pi}{3}. \tag{8.2}$$

The d-q plane is divided into six 60°-wide sectors, designated 1 through 6, and centered on the corresponding voltage vectors (notice that these sectors are different from these in Figure 4.23). A stator flux vector, $\lambda_s = \lambda_s \exp(j\Theta_s)$, is said to be associated with the voltage vector v_K when

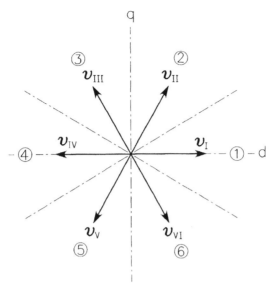

FIGURE 8.3   Space vectors of the inverter output voltage and sectors of the vector plane.

it passes through Sector K, which means that of all the six voltage vectors, the orientation of v_K is closest to that of λ_s. For example, the stator flux vector becomes associated with v_{II} when passing through Sector 2. In another example, when a phase of the same vector is 200°, then it is associated with the voltage vector v_{IV}.

Impacts of individual voltage vectors on the stator flux and developed torque, when λ_s is associated with v_K, are listed in Table 8.1. The impact of vectors v_K and v_{K+3} on the developed torque is ambiguous, because it depends on whether the flux vector is leading or lagging the voltage vector in question. The zero vector, v_Z, that is, v_0 or v_7, does not affect the flux but reduces the torque, because the vector of rotor flux gains on the stopped stator flux vector.

A block diagram of the classic DTC drive system is shown in Figure 8.4. The dc-link voltage (which, although supposedly constant, tends to fluctuate a little), V_i, and two stator currents, i_a and i_b, are measured, and

TABLE 8.1   Impact of Individual Voltage Vectors on the Stator Flux and Developed Torque

	v_K	v_{K+1}	v_{K+2}	v_{K+3}	v_{K+4}	v_{K+5}	v_Z
λ_s	↗↗	↗	↘	↘↘	↘	↗	-
T_M	?	↗	↗	?	↘	↘	↘

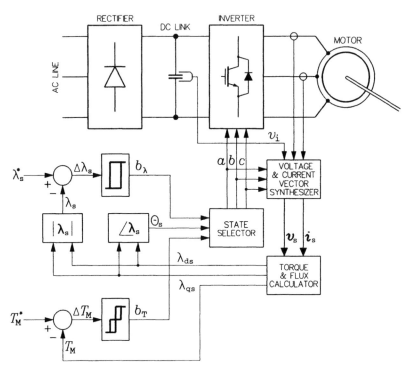

FIGURE 8.4 Block diagram of the DTC drive system.

space vectors, v_s^s and i_s^s, of the stator voltage and current are determined in the voltage and current vector synthesizer. The voltage vector is synthesized from V_i and switching variables, a, b, and c, of the inverter, using Eq. (4.3) or (4.8), depending on the connection (delta or wye) of stator windings. Based on v_s and i_s, the stator flux vector, λ_s, and developed torque, T_M, are calculated. The magnitude, λ_s, of the stator flux is compared in the flux control loop with the reference value, λ_s^*, and T_M is compared with the reference torque, T_M^*, in the torque control loop.

The flux and torque errors, $\Delta\lambda_s$ and ΔT_M, are applied to respective bang-bang controllers, whose characteristics are shown in Figure 8.5. The flux controller's output signal, b_λ, can assume the values of 0 and 1, and that, b_T, of the torque controller can assume the values of -1, 0, and 1. Selection of the inverter state is based on values of b_λ and b_T. It also depends on the sector of vector plane in which the stator flux vector, λ_s, is currently located (see Figure 8.3), that is, on the angle Θ_s, as well as on the direction of rotation of the motor. Specifics of the inverter state selection are provided in Table 8.2 and illustrated in Figure 8.6 for the stator flux vector in Sector 2. Five cases are distinguished: (1) Both the

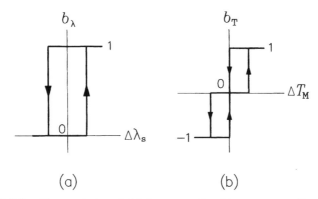

FIGURE 8.5   Characteristics of: (a) flux controller, (b) torque controller.

TABLE 8.2   Selection of the Inverter State in the DTC Scheme; (a) Counterclockwise Rotation

b_λ	1			0		
b_T	1	0	−1	1	0	−1
Sector 1	6	7	5	2	0	1
Sector 2	2	0	4	3	7	5
Sector 3	3	7	6	1	0	4
Sector 4	1	0	2	5	7	6
Sector 5	5	7	3	4	0	2
Sector 6	4	0	1	6	7	3

(b) Clockwise Rotation

b_λ	1					0
b_T	1	0	−1	1	0	−1
Sector 1	5	7	6	1	0	2
Sector 2	1	0	4	3	7	6
Sector 3	3	7	5	2	0	4
Sector 4	2	0	1	6	7	5
Sector 5	6	7	3	4	0	1
Sector 6	4	0	2	5	7	3

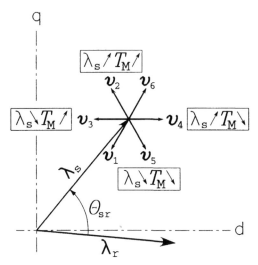

FIGURE 8.6 Illustration of the principles of inverter state selection.

flux and torque are to be decreased; (2) the flux is to be decreased, but the torque is to be increased; (3) the flux is to be increased, but the torque is to be decreased; (4) both the flux and torque are to be increased; and (5) the torque error is within the tolerance range. In Cases (1) to (4), appropriate nonzero states are imposed, while Case (5) calls for such a zero state that minimizes the number of switchings (State 0 follows States 1, 2, and 4, and State 7 follows States 3, 5, and 6).

EXAMPLE 8.1 The inverter feeding a counterclockwise rotating motor in a DTC ASD is in State 4. The stator flux is too high, and the developed torque is too low, both control errors exceeding their tolerance ranges. With the angular position of stator flux vector of 130°, what will be the next state of the inverter? Repeat the problem if the torque error is tolerable.

In the first case, the output signals of the flux and torque controllers are $b_\lambda = 0$ and $b_T = 1$. The stator flux vector, λ_s, is in Sector 3 of the vector plane. Thus, according to Table 8.2, the inverter will be switched to State 1. In the second case, $b_T = 0$, and State 0 is imposed, by changing the switching variable a from 1 to 0. ■

To illustrate the impact of the flux tolerance band on the trajectory of λ_s, a wide and a narrow band are considered, with b_T assumed to be 1. The corresponding example trajectories are shown in Figure 8.7. Links between the inverter voltage vectors and segments of the flux trajectory are also indicated. Similarly to the case of current control with hysteresis

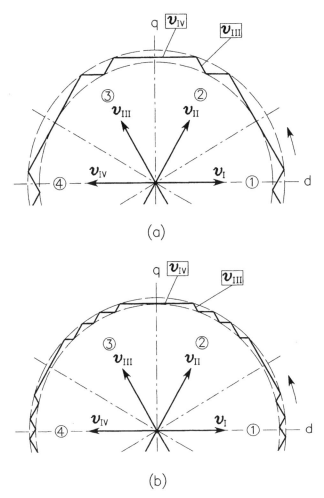

FIGURE 8.7   Example trajectories of the stator flux vector ($b_T = 1$): (a) wide error tolerance band, (b) narrow error tolerance band.

controllers (see Section 4.5), the switching frequency and quality of the flux waveforms increase when the width of the tolerance band is decreased.

The only parameter of the motor required in the DTC algorithm is the stator resistance, R_s, whose accurate knowledge is crucial for high-performance low-speed operation of the drive. Low speeds are accompanied by a low stator voltage (the CVH principle is satisfied in all ac ASDs), which is comparable with the voltage drop across R_s. Therefore, modern DTC ASDs are equipped with estimators or observers of that resistance. Various other improvements of the basic scheme described, often involving machine intelligence systems, are also used to improve the dynamics and efficiency of the drive and to enhance the quality of stator currents in the motor. An interesting example of such an improvement is the "sector shifting" concept, employed for reducing the response time of the drive to the torque command. It is worth mentioning that this time is often used as a major indicator of quality of the dynamic performance of an ASD.

As illustrated in Figure 8.8, a vector of inverter voltage used in one sector of the vector plane to *decrease* the stator flux is employed in the next sector when the flux is to be *increased*. With such a control and with the normal division of the vector plane into six equal sectors, the trajectory of stator flux vector forms a piecewise-linear approximation of a circle. Figure 8.9 depicts a situation in which, following a rapid change in the torque command, the line separating Sectors 2 and 3 is shifted back by α radians. It can be seen that the inverter is "cheated" into applying vectors v_V and v_{IV} instead of v_{IV} and v_{III}, respectively. Note that the linear speed of travel of the stator flux vector along its trajectory is constant and equals the dc supply voltage of the inverter. Therefore, as that vector takes now a "shortcut," it arrives at a new location in a shorter time than if it traveled along the regular trajectory. The acceleration of stator flux vector described results in a rapid increase of the torque, because that vector quickly moves away from the rotor flux vector. The greater the sector shift, α, the greater the torque increase. It can easily be shown (the reader is encouraged to do that) that expanding a sector, that is shifting its border forward ($\alpha < 0$), leads to instability as the flux vector is directed toward the outside of the tolerance band.

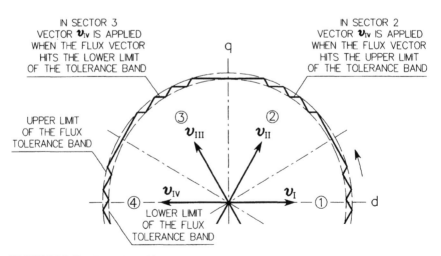

FIGURE 8.8 Selection of inverter voltage vectors under regular operating conditions.

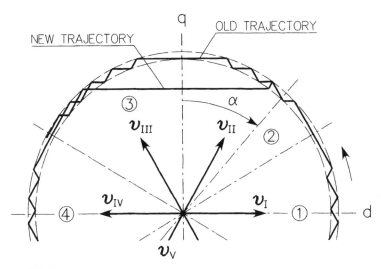

FIGURE 8.9   Acceleration of the stator flux vector by sector shifting.

To highlight the basic differences between the direct field orientation (DFO), indirect field orientation (IFO), and DTC schemes, general block diagrams of the respective drive systems are shown in Figures 8.10 to 8.12. The approach to inverter control in the DFO and IFO drives is distinctly different from that in the DTC system. Also, the bang-bang hysteresis controllers in the latter drive contrast with the linear flux and torque controllers used in the field orientation schemes.

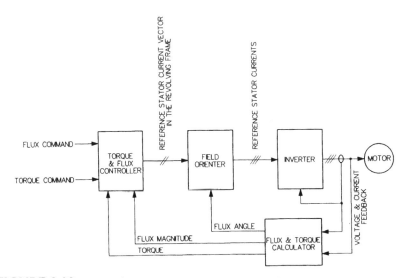

FIGURE 8.10   Simplified block diagram of the direct field orientation scheme.

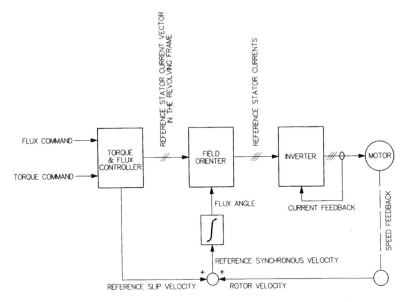

FIGURE 8.11 Simplified block diagram of the indirect field orientation scheme.

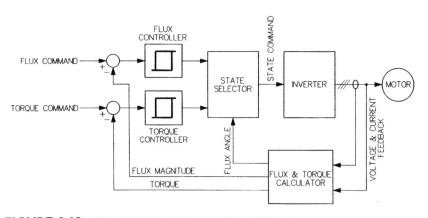

FIGURE 8.12 Simplified block diagram of the DTC scheme.

8.3 DIRECT SELF-CONTROL

The Direct Self-Control (DSC) method, proposed by Depenbrock in 1985, is intended mainly for high-power ASDs with voltage source inverters. Typically, slow switches, such as GTOs, are employed in such inverters, and low switching frequencies are required. Therefore, in DSC drives, the inverter is made to operate in a mode similar to the square-wave one, with occasional zero states thrown in. The zero states disappear when the drive runs with the speed higher than rated, that is, in the field weakening area, where, as in all other ASDs, the inverter operates in the square-wave mode.

DSC ASDs are often misrepresented as a subclass of DTC drives. However, the principle of DSC is different from that of DTC. To explain this principle, note that while the output voltage waveforms in voltage source inverters are discontinuous, the time integrals of these waveforms are continuous and, in a piecewise manner, they approach sine waves. It can be shown that using these integrals, commonly called *virtual fluxes*, and hysteresis relays in a feedback arrangement, the square-wave operation of the inverter may be enforced with no external signals (hence the "self" term in the name of the method). The output frequency, f_o, of the so-operated inverter is proportional to the V_i/λ^* ratio, where V_i denotes the dc input voltage of the inverter and λ^* is the reference magnitude of the virtual flux. Specifically,

$$f_o = \frac{1}{4\sqrt{3}} \frac{V_i}{\lambda^*} \tag{8.4}$$

when the virtual fluxes are calculated as time integrals of the line-to-line output voltages of the inverter, and

$$f_o = \frac{1}{6} \frac{V_i}{\lambda^*} \tag{8.5}$$

when the line-to-neutral voltages are integrated. The self-control scheme is illustrated in Figure 8.13 and characteristics of the hysteresis relays are shown in Figure 8.14. Waveforms of the virtual fluxes are depicted in Figure 8.15 (notice the negative phase sequence). As shown in Figure 8.16, the trajectories of the corresponding flux vectors, λ, are hexagonal, both for the line-to-neutral and line-to-line voltage integrals. As in the case of motor variables, the magnitude, λ, of the flux vector is 1.5 times greater than the amplitude of flux waveforms.

In spite of the hexagonal trajectory and nonsinusoidal waveforms of virtual fluxes, the total harmonic distortion of these waveforms is low. It can further be reduced by the so-called *corner folding*, illustrated in Figure 8.17. The flux trajectory becomes closer to a circle, albeit at the expense of a somewhat more complicated self-control scheme and a threefold increase in the switching frequency.

EXAMPLE 8.2 What is the total harmonic distortion of the trapezoidal waveform of virtual flux depicted in Figure 8.15(b)?

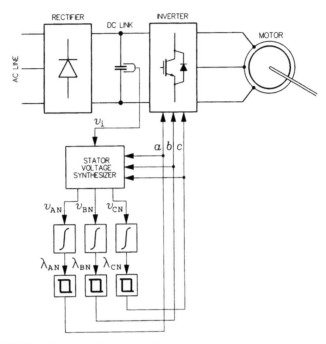

FIGURE 8.13 Inverter self-control scheme.

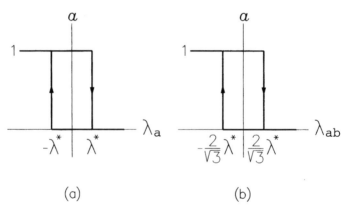

FIGURE 8.14 Characteristics of the hysteresis relays in the inverter self-control scheme: (a) line-to-neutral voltages integrated, (b) line-to-line voltages integrated.

The total harmonic distortion, THD, of the waveform in question is defined as the ratio of the harmonic content, Λ_h, of this waveform to the fundamental flux, Λ_1. The harmonic content can be determined as a geometrical difference (square root of a difference of squares) of the rms value, Λ, and the fundamental, Λ_1.

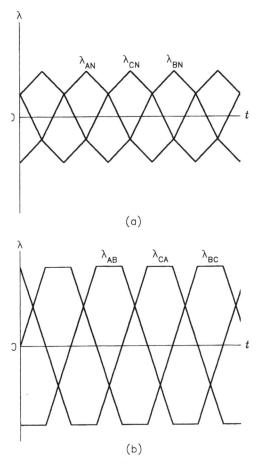

FIGURE 8.15   Waveforms of virtual fluxes: (a) line-to-neutral voltage integrals, (b) line-to-line voltage integrals.

The virtual flux waveform has the odd symmetry and half-wave symmetry. Therefore, it is sufficient to consider a half of it, from 0° to 180°. Denoting the peak value of the waveform by Λ_{max}, the expression for the waveform is

$$\lambda(\omega t) = \begin{cases} \dfrac{3}{\pi}\Lambda_{max}\omega t & for \quad 0 \leq \omega t < \dfrac{1}{3}\pi \\[2mm] \Lambda_{max} & for \quad \dfrac{1}{3}\pi \leq \omega t < \dfrac{2}{3}\pi. \\[2mm] \dfrac{3}{\pi}\Lambda_{max}(\pi - \omega t) & for \quad \dfrac{2}{3}\pi \leq \omega t < \pi \end{cases}$$

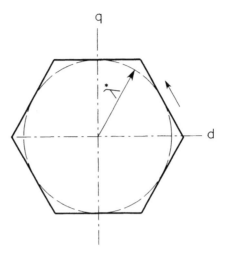

FIGURE 8.16   Hexagonal trajectory of the virtual flux vector.

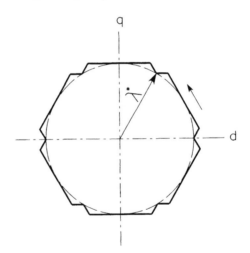

FIGURE 8.17   Trajectory of the virtual flux vector with "corner folding."

The rms value of the virtual flux, calculated as

$$\Lambda = \sqrt{\frac{1}{\pi}\int_0^\pi \lambda^2(\omega t)d\omega t},$$

is $\sqrt{5}\Lambda_{max}/3 \approx 0.7454\Lambda_{max}$, and the rms value of fundamental flux, given by

$$\Lambda_1 = \frac{\sqrt{2}}{\pi}\int_0^\pi \lambda(\omega t)\sin(\omega t)d\omega t,$$

is $3\sqrt{6}\Lambda_{max}/\pi^2 \approx 0.7446\Lambda_{max}$. Thus, the harmonic content of the virtual flux waveform is

$$\Lambda_h = \sqrt{\frac{5}{9} - \frac{54}{\pi^4}}\Lambda_{max} \approx 0.0345\Lambda_{max},$$

and the total harmonic distortion is

$$THD = \frac{\Lambda_h}{\Lambda_1} = \frac{0.0345\Lambda_{max}}{0.7446\Lambda_{max}} \approx 0.046;$$

that is, less than 5%. ∎

The angular velocity of the virtual flux vector can be changed by: (a) halting the vector by switching the inverter to a zero state, (b) applying a voltage vector that will attempt to move the stator flux vector in the direction opposite to that of regular rotation, or (c) changing the reference magnitude, λ^* [see Eqs. (8.4) and (8.5)]. The latter approach is illustrated in Figure 8.18. The reference magnitude of the flux vector is λ_1^*, but when the vector reaches point A on its trajectory, the reference magnitude is temporarily reduced to λ_2^*. As a result, instead of passing through points B and C, the flux vector takes a shortcut to point D. Because, as already mentioned in Section 8.2, the linear speed of travel of the flux vector is constant and the AD trajectory portion is shorter than the ABCD portion, the flux vector arrives in point D sooner than if it had followed the original trajectory. The "trajectory deformation" effect described is similar to that

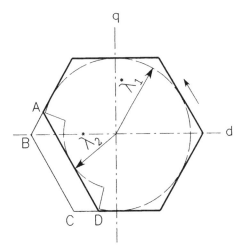

FIGURE 8.18 Acceleration of the virtual flux vector by reduction of the reference magnitude of the vector.

caused by sector shifting in the DTC technique (see Figure 8.9), but here it is accomplished by a different means. Under the field weakening conditions, the flux is reduced permanently. Note that these conditions are associated with high speeds of the motor and its vectors. Thus, the field weakening feature is inherent for the DSC control scheme.

In practical DSC drives, stator EMFs are integrated instead of stator voltages, so that the virtual flux vector becomes a stator flux vector. A block diagram of the classic DSC drive is shown in Figure 8.19. The stator EMF vector, $\mathbf{e_s}$, and developed torque, T_M, are determined on the basis of stator voltage and current vectors. Components, e_{ds} and e_{qs}, of the EMF vector are integrated to reproduce components, λ_{ds} and λ_{qs}, of the stator flux vector, which are later converted to the estimated stator flux waveforms, λ_a, λ_b, and λ_c, for individual phases of the motor. These are applied to hysteresis relays similar to those in the inverter self-control scheme in Figure 8.13. The hysteresis width of the relays is controlled by the flux reference signal, λ_s^* (see Figure 8.14). Resultant reference values, a^*, b^*, and c^*, of switching variables of the inverter feeding the motor and the output signal, b_T, from the closed-loop torque control circuit

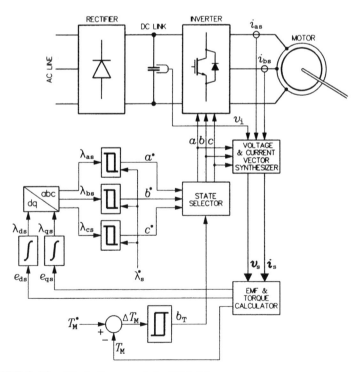

FIGURE 8.19 Block diagram of the DSC drive system.

are forwarded to a state selector. The selector generates switching signals, a, b, and c, which, if expressing active states of the inverter, equal their respective reference values. However, if a zero state is needed, the selector chooses State 0 or State 7, based on the previous state of the inverter, so that switches in only one inverter phase are activated. In more advanced versions of the DSC schemes, the features of corner folding and trajectory deformation and other enhancements are employed.

8.4 SPACE-VECTOR DIRECT TORQUE AND FLUX CONTROL

Over the years, many improvements have been added to the basic version of the DTC method described in Section 8.2, including more robust and accurate calculators of the controlled variables. The DTC consists in the bang-bang control of the torque and flux, being thus characterized by a fast response to control commands. However, in the steady state, the DTC principle results in chaotic switching patterns in the inverter. Unless a high average switching frequency is enforced by setting the hysteresis loops of the torque and flux controllers to low values (see Figures 8.4 and 8.5), significant torque ripple and undesirable acoustic and vibration effects associated with that ripple are produced. Therefore, one of the main goals of the aforementioned improvements is to optimize the steady-state switching process.

The popular space-vector PWM strategy for voltage-source inverters, presented in Section 4.5, is widely recognized as a means for generating high-quality switching patterns. Such patterns result in low-ripple stator currents and, consequently, smooth flux and torque waveforms. Therefore, a class of DTC techniques has been developed in which the control system generates a reference vector of stator voltage, instead of directly indicating the next state of the inverter. The reference voltage vector is then realized using the space-vector pulse width modulation. Description of an example space-vector direct torque and flux control system follows.

Consider Eq. (6.27) with the revolving reference frame aligned with the stator flux vector, that is, $\lambda_{DS} = \lambda_s$, $\lambda_{QS} = 0$. Then, the real and imaginary part of that equation may be written as

$$v_{DS} = R_s i_{DS} + \frac{d\lambda_s}{dt} \tag{8.6}$$

and

$$v_{QS} = R_s i_{QS} + \omega_e \lambda_s. \tag{8.7}$$

The developed torque, T_M, is given by Eq. (7.4), and Eq. (8.7) can be rewritten as

$$v_{QS} = \frac{1.5}{P_p} R_s \frac{T_M}{\lambda_s} + \omega_e \lambda_s. \tag{8.8}$$

The angular velocity, ω_e, of the stator flux vector appearing in the last relation may be estimated from the last two stator flux vector signals, $\lambda_{s(k)}$ and $\lambda_{s(k-1)}$, as

$$\omega_e = \frac{\lambda_{QS(k)}\lambda_{DS(k-1)} - \lambda_{DS(k)}\lambda_{QS(k-1)}}{(\lambda_{DS(k)}^2 + \lambda_{QS(k)}^2)\tau_{smp}}, \tag{8.9}$$

where τ_{smp} denotes the sampling period of the digital control system.

It can be seen from Eq. (8.6) that the V_{DS} component of stator voltage vector has a strong impact on the rate of change of stator flux. In turn, according to Eq. (8.8), the V_{QS} component can be used for control of the developed torque. These observations underlie the operating principle of the drive system under consideration. A block diagram of the core of that system is shown in Figure 8.20. Linear PI controllers are used in the flux and torque control loops. In the latter loop, the signal from torque controller is augmented with the $\omega_e \lambda_s$ signal (see Eq. 8.9) to improve the dynamic response to the torque command by decoupling the two control channels.

The angular position, Θ_s, of the stator flux vector is required for conversion of the reference stator voltage vector, $v_s^{e*} = v_{DS}^* + jv_{QS}^*$, in the revolving reference frame into the same vector, $v_s^{s*} = v_{ds}^* + jv_{qs}^*$, in the stator frame. It is obtained, together with λ_s, from a stator flux calculator (estimator or observer). Analogously, the torque signal, T_M, comes from

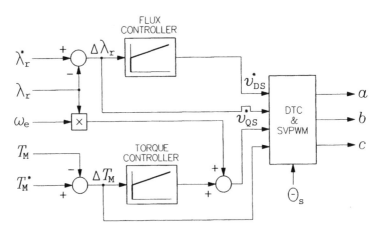

FIGURE 8.20   Block diagram of the core of SVDTC drive system.

a torque calculator. The reference vector, $v^* = v_s^{s*}$, of the inverter output voltage allows determination of the switching signals, a, b, and c, using the space-vector PWM method.

If a sudden change in control commands occurs, the reference voltage vector may exceed the output voltage capability of the inverter. Therefore, in such situations, the inverter control is switched to the classic DTC scheme, which results in a fast response to those commands. Thus, in the steady state, the inverter operates with the space-vector pulse width modulation, while under transient conditions it is controlled by the bang-bang controllers of torque and flux.

8.5 SUMMARY

Direct torque and flux control methods consist in selection of consecutive states of the inverter feeding the induction motor. Although the stator and rotor flux vectors are similar with respect to the magnitude and angular position, they differ significantly in their dynamic response to changes in the stator voltage. Specifically, reaction of the rotor flux vector is more sluggish than that of the vector of stator flux. Therefore, application of appropriate voltage vectors, associated with individual inverter states, allows adjusting the magnitude of stator flux as well as manipulating the angle between the stator and rotor flux vectors. Because the torque developed in the motor is proportional to the sine of that angle, the magnetic field and torque of the motor can simultaneously be controlled.

Bang-bang controllers of the flux and torque are employed in the Direct Torque Control (DTC) drives, while integrators of stator EMFs are used in the feedback loop of the Direct Self-Control (DSC) scheme, designated for high-power ASDs. To improve the steady-state operation of DTC systems, the reference space vector of stator voltage is determined and realized using the space-vector PWM technique.

9

SPEED AND POSITION CONTROL

To begin this chapter, variables controlled in induction motor drives are specified; then principles of machine intelligence controllers are explained. Chapter 9 concludes with a description of speed and position control systems with linear, variable structure, and machine intelligence controllers.

9.1 VARIABLES CONTROLLED IN INDUCTION MOTOR DRIVES

As explained in Chapter 5, scalar control methods for induction motor drives allow control of the steady-state speed or torque, with the intensity of magnetic field in the motor maintained at an approximately constant level. Vector control techniques, described in Chapters 7 and 8, are designed to independently adjust the developed torque and selected magnetic flux, both in the steady state and under transient operating conditions of the drive. Also, as demonstrated in Section 7.2, the torque in a field-oriented motor is produced under the optimal conditions of orthogonality of the flux and current vectors, yielding the maximum possible torque-per-ampere ratio.

The ability of adjusting the strength of magnetic field in the motor is utilized in loss-minimization algorithms. With high loads, the optimal strategy is to keep the flux at the rated level, but with low loads, often occurring in practical drives, the best trade-off between copper and iron losses can be obtained by proper reduction of the flux. When a motor is run with a speed higher than rated, the flux is reduced too (field weakening), but for a different reason. Maintaining the flux at the rated level would require increasing the stator voltage above the rated value, which might damage the insulation of stator windings.

Autonomous control of the developed torque is necessary in winder drives, in which at least two drive systems must cooperate to maintain the desired tension and linear speed of the wound tape. Such an arrangement is illustrated in Figure 9.1. To control the tension, F, and speed, u, of the tape, the speed, ω_M, of Motor 1 is adjusted in inverse proportion to the radius, r_1, of Coil 1, while the torque, T_M, developed in Motor 2 is controlled in direct proportion to the radius, r_2, of Coil 2. Thus, Motor 1 is speed controlled and Motor 2 is torque controlled. This is indicated in the figure in question by appropriate placement of symbols ω_M^* and T_M^*, denoting the reference speed for Motor 1 and reference torque for Motor 2, respectively.

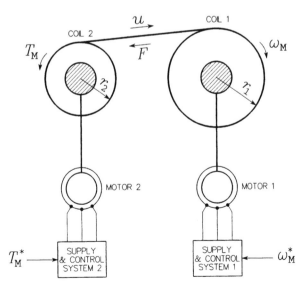

FIGURE 9.1   Winder drive system.

Electric cars, in some of which ASDs with induction motors are employed, present another example of autonomous torque control. In order to emulate driving characteristics of gasoline-powered automobiles, the accelerator pedal provides the reference torque signal. As in traditional cars, in the cruise-control mode, the electric car is speed controlled using a closed-loop control scheme.

In most ASDs, the speed and, less often, the angular displacement of the rotor are the principal variables controlled, with the torque and flux controls subordinated to the speed or position control algorithms. Controlled-speed, or *spindle*, drive systems are most common in practice while, as already mentioned in Chapter 1, positioning, or *servo*, drive systems are the most sophisticated.

9.2 SPEED CONTROL

A general block diagram of a high-performance induction motor drive with speed control is shown in Figure 9.2. The speed of the motor, ω_M, is measured (or estimated) and compared with the reference signal, ω_M^*. Usually the speed is determined as a time derivative of the rotor angle, Θ_M. The speed error, $\Delta\omega_M$, and ω_M^* signals are forwarded to the speed

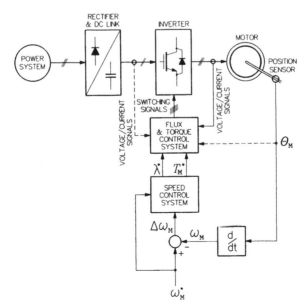

FIGURE 9.2   General block diagram of a high-performance adjustable speed drive.

control system, which produces reference values, λ_f^* and T_M^*, of the flux (stator, rotor, or air-gap) and torque of the motor. These, with other applicable signals from various parts of the drive, are employed in the flux and torque control system based on the FO, DTC, or DSC principle.

In most practical ASDs, the flux is controlled according to the formula

$$\lambda_f^* = \begin{cases} \lambda_{f,rat} & \text{for} \quad \omega_M^* \leq \omega_{M,rat} \\ \dfrac{\omega_{M,rat}}{\omega_M^*}\lambda_{f,rat} & \text{for} \quad \omega_M^* > \omega_{M,rat} \end{cases}, \qquad (9.1)$$

where $\lambda_{f,rat}$ and $\omega_{M,rat}$ denote rated values of the controlled flux and motor velocity, respectively. Eq. (9.1) is equivalent to Eq. (5.25), ensuring that the stator voltage under the field weakening conditions will not exceed the rated value. A block diagram of such a flux control scheme is shown in Figure 9.3. In drives operating most of the time with low load torque, loss-minimizing algorithms can be used to control the flux. Either it is adjusted dependent on that torque, or an automated search is performed to determine the minimum-loss operating point. As already mentioned in Section 9.1, such algorithms aim at the optimal trade-off between iron and copper losses.

The classic approach to torque control in speed-controlled drives, illustrated in Figure 9.4, utilizes a linear speed controller producing the reference value, T_M^*, of the motor torque. Linear speed controllers are, typically, of the proportional-integral (PI) type, that is,

$$T_M^* = k_P\Delta\omega_M + k_I\int_0^t \Delta\omega_M dt, \qquad (9.2)$$

where k_P and k_I denote the proportional and integral gains of the controller, respectively. Linear controllers represent a mature technology of the feed-

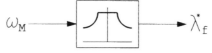

FIGURE 9.3   Flux control scheme.

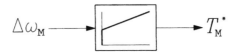

FIGURE 9.4   Torque control scheme.

back control and are quite effective. However, they are not free from certain weaknesses, such as the need for proper selection of coefficients k_P and k_I or the nonoptimal dynamic response to rapid changes in the speed command. Therefore, several alternatives to these controllers have been developed over the years, including use of such old technologies as the phase-locked loop (PLL), dating back to early 1930s (see the survey paper by Hsieh and Hung, 1996).

Robustness and swiftness of the speed control can greatly be enhanced employing a *variable structure control* (VSC) technique, which is based on the idea of the so-called *sliding mode*. Eschewing the mathematics of VSC, it can simply be explained as forcing the controlled variable to stay within a prescribed tolerance band by reacting strongly and immediately when the variable departs from the band. A case of such control, using hysteresis controllers, was already encountered in Section 4.5 with respect to output currents in voltage source inverters. Two basic realizations of a variable structure speed controller for high-performance ASDs are illustrated in Figure 9.5.

The relay controller shown in Figure 9.5(a) generates one of two values of the reference torque, T_M^*, depending on the sign and value of the speed error, $\Delta\omega_M$. When the error is so high that it exceeds the upper value of a tolerance band set up in the controller, T_M^* is made equal to T, a high, but feasible, value of the motor torque. Conversely, with the

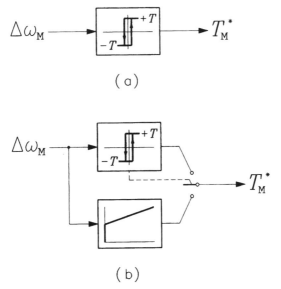

(a)

(b)

FIGURE 9.5 Variable-structure speed controllers: (a) relay, (b) relay/linear.

speed error too low, the reference torque is set to $-T$. This control algorithm, although very effective in keeping the speed error within the tolerance band, results in "chattering," that is, a high-frequency ripple of the speed, ω_M. Note that with the error at a tolerable level, the reference torque is still at one of its extremes, unnecessarily forcing the speed error out of the tolerance band. A simple solution, shown in Figure 9.5(b), consists of using a parallel linear controller, which provides the reference torque when $\Delta\omega_M$ stays within the tolerance limits.

In recent years, controllers based on principles of *machine intelligence* (MI) have increasingly been employed in advanced ASDs. Machine intelligence systems try to emulate the human brain, and, thanks to their robustness and adaptivity, they are particularly effective in applications characterized by complex or poorly defined mathematical models. Three types of MI controllers stand out: (a) neural controllers, (b) fuzzy controllers, and (c) neurofuzzy controllers, combining operating principles of the neural and fuzzy controllers. Because they are likely to dominate control systems of the future, the whole next section is devoted to them.

9.3 MACHINE INTELLIGENCE CONTROLLERS

The most common *feedforward neural network* is shown in Figure 9.6. The circles represent basic components of the network, called *neurons*. The network consists of three *layers*: the input layer, the hidden layer, and the output layer. Each of the k inputs is connected to each of m

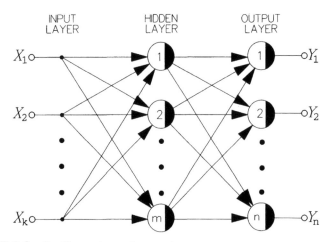

FIGURE 9.6 Feedforward neural network.

neurons in the hidden layer and, in turn, each of the hidden-layer neurons is connected to each of n output neurons. The arrows represent *connection weights* (multipliers), the signal at the end of an arrow being obtained by multiplying the signal at the beginning of the arrow by the weight. The whole network has k *inputs*, X_1, X_2, ..., X_k, and n *outputs*, Y_1, Y_2, ..., Y_n. A single neuron, depicted in Figure 9.7, consists of an adder and a nonlinear *squashing function*, $y = f(s)$. Typically, a sigmoidal squashing function is used, that is,

$$ y = \frac{1}{1 + e^{-Ks}}, \tag{9.3} $$

where K is a constant. If $-\infty < s < \infty$, then $0 < y < 1$, and if K approaches infinity, the function approaches the unit step function. Nonlinearity of the squashing function allows the neural network to model nonlinear phenomena. The argument, s, of the squashing function is given by

$$ s = b + \sum_{i=1}^{j} c_i x_i, \tag{9.4} $$

where x_i, $i = 1, 2, ..., j$, are neuron's input signals, c_i are the respective connection weights, and b is the *bias weight*.

To work properly, a neural network must be *trained*, that is, its weights must be set to values resulting in the required quality of operation. The training process consists of sequential application of various sets of inputs, evaluating the difference between the outputs and their reference values, and adjusting the weights to minimize that difference. In neural networks of the type described, the training process is usually based on the efficient *backpropagation* algorithm. The training is automated and performed prior to employing the network in an application. Often it is permanently continued under regular operating conditions.

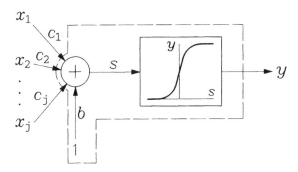

FIGURE 9.7 Single neuron.

Neural networks, an ingenious idea of signal processing, can be implemented in various ways. Software neural networks are simply computer programs realized in microcomputers or digital signal processors (DSPs). Operating in the sequential mode, they emulate the parallel processing of data, characteristic for the ideal neural network. Hardware networks, still mostly in the developmental stage, are dedicated analog or digital integrated circuits with true parallelism of operation.

In ASDs, neural networks have been used as current controllers, with current errors as inputs and the inverter switching signals as outputs. In inverters with feedforward voltage control, neural networks can be employed for the generation of optimal switching patterns in dependence on the modulation index. A neural network with input signals representing the dc-link current (as an indirect measure of the motor load) and reference speed of the motor can be trained as a loss-minimizing source of the reference flux values, λ_f^*. Also, as in many other applications, neural networks are an excellent tool for automated fault detection and diagnosis.

Fuzzy controllers work similarly to human operators, who seldom process strictly numerical data. Instead, humans tend to issue and follow such imprecise control commands as "more," "slowly," or "a bit to the left." The underlying theory, the *fuzzy logic*, disposes of the unambiguous assignment of the "true" or "false" values to its variables, typical for the classic, "crisp" logic. Instead, a fuzzy variable is said to have a certain probability, P, of being "true" and the complementary probability, $1 - P$, of being "false." With respect to signals in fuzzy control systems, *fuzzy* (linguistic) values, such as "low," "medium," or "high," are employed. Each fuzzy value represents a fuzzy set, which entails a specified range of *crisp* (numerical) values. Degree of belongingness of each crisp value in the set, called a *membership grade*, is given by a *membership function*, μ. Various membership functions are used, such as triangular, trapezoidal, or bell-shaped.

The concept of fuzzy variables and membership functions is illustrated in Figure 9.8 for the torque, T_M, of an example motor. The whole torque range is divided into six overlapping fuzzy sets of numerical values. To each of these sets, a linguistic value is assigned, namely the "negative high" (NH), "negative medium" (NM), "negative low" (NL), "positive low" (PL), "positive medium" (PM), and "positive high" (PH). Trapezoidal membership functions, μ_T, are used. It can be seen, for instance, that a torque of -5 Nm is defined as purely "negative low." However, a torque of 12 Nm belongs in both the "positive low" and "positive medium" sets. Its membership grade is 0.8 in the PL set and 0.2 in the PM set.

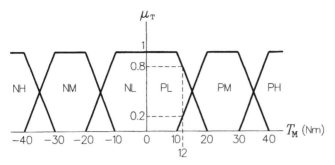

FIGURE 9.8   Fuzzy sets of torque values of an example motor.

A typical fuzzy control system is illustrated in Figure 9.9. The (C) and (F) symbols represent crisp and fuzzy values, respectively. Raw output signals from the controlled plant are preprocessed and applied to the *fuzzifier*. The preprocessing includes determination of control errors, and it may involve calculation of these controlled variables, which cannot be measured directly. Based on predefined membership functions, the fuzzifier assigns one or more of fuzzy values to each crisp variable received. The resultant fuzzy variables, with their membership grades, are then forwarded to the *inference engine*, the heart of the fuzzy controller. The membership grades can be thought of as "weights" of individual fuzzy inputs to the inference engine.

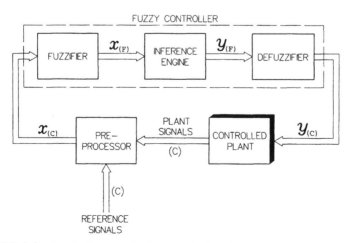

FIGURE 9.9   Block diagram of a fuzzy control system.

The inference engine is based on a set of expert rules, of the conditional "if-then" format. The outcome of a rule, y, can be fuzzy, as in

$$\text{if } x_1 \text{ is } A \text{ and } x_2 \text{ is } B, \text{ then } y \text{ is } C, \qquad (9.5)$$

where x_1 and x_2 are inputs to the fuzzy controller, and A, B, and C are linguistic values, or crisp, as in

$$\text{if } x_1 \text{ is } A \text{ and } x_2 \text{ is } B, \text{ then } y = f(x_1, x_2), \qquad (9.6)$$

where A, B, and C are linguistic values and $f(x_1, x_2)$ denotes a given function of the inputs. The rules, derived from a *knowledge base*, represent the best decisions regarding the control action in response to control errors and other signals from the plant. As typical for most control methods, the main objectives of fuzzy control are: (a) minimizing the control errors, (b) limiting system variables to allowable ranges, and (c) avoiding unnecessary action if the control errors remain within predefined tolerance bands.

The aggregated output set of the inference engine is a combination of weighted fuzzy sets resulting from individual expert rules. The weight of a rule, formally called a *firing strength*, is calculated from membership grades of the inputs involved. With several inputs, the weight of the rule can, for instance, be taken as a product of their individual membership grades. The last part of the fuzzy controller, the *defuzzifier*, extracts from the output set a crisp value of the control signal for the controlled plant. In general, the fuzzy controller can be multidimensional, generating several control signals in a parallel fashion. Several defuzzification methods have been developed, the most common, so-called *centroid* technique consisting in determination of the gravity center of output set of the inference engine.

EXAMPLE 9.1 For simplicity, a single-input single-output fuzzy controller is considered. Fuzzy sets of the input and output variables, x and y, are shown in Figures 9.10(a) and 9.10(b), respectively. The expert rules used by the inference engine are:

Rule 1: If x is "low," then set y to "high."
Rule 2: If x is "medium," then set y to "medium."
Rule 3: If x is "high," then set y to "low."

Find the crisp value, $y_{(C)}$, of the output variable if $x_{(C)} = 1.25$.

As seen from Figure 9.10(a), the considered value of x is both "low," with the membership grade of 0.75, and "medium," with the grade of 0.25. Consequently, Rules 1 and 2 apply, yielding $y = $ "high" and $y = $ "medium," respectively. Firing strengths of these rules are

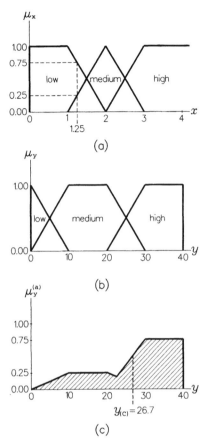

FIGURE 9.10   Illustration of Example 9.1: (a) fuzzy sets of the input variable, (b) fuzzy sets of the output variable, (c) weighted output set.

0.75 and 0.25, respectively. The aggregated output fuzzy set, shown in Fig. 9.10(c), is thus constructed as a *union* (sum) of the $y =$ "high" set multiplied, heightwise, by 0.75 and the $y =$ "medium" set multiplied by 0.25. The crisp value, $y_{(C)}$, of the output of the fuzzy controller is determined as the y-coordinate of the centroid (center of gravity) of the output set. It is calculated as

$$y_{(C)} = \frac{\int\limits_{0}^{40} \mu_y^{(a)} y \, dy}{\int\limits_{0}^{40} \mu_y^{(a)} dy}$$

and equals 26.7.   ∎

It can be seen that the setup of a fuzzy controller is quite arbitrary with respect to selection of the fuzzy values and assignment of the corresponding membership functions. The expert rules are, generally, imprecise. That is why the fuzzy control may be not the best solution if the mathematical model of controlled plant can accurately be determined. This, however, is not a common situation in practice as, in the words of the physicist Augustin Fresnel, "Nature is not embarrassed by difficulties in analysis."

Neurofuzzy controllers have been developed to enhance the adaptivity of fuzzy controllers by combining the principles of fuzzy control and neural networks. The ANFIS (adaptive neurofuzzy inference system) networks stand out in this class of MI controllers. In the subsequent considerations, two inputs, x_1 and x_2, and one output, y, of the ANFIS controller are assumed. A typical rule set with two fuzzy "if-then" rules is

$$\text{if } x_1 \text{ is } A_1 \text{ and } x_2 \text{ is } B_1, \text{ then } y = f_1(x_1,x_2) = a_1x_1 + b_1x_2 + c_1 \tag{9.7}$$

and

$$\text{if } x_1 \text{ is } A_2 \text{ and } x_2 \text{ is } B_2, \text{ then } y = f_2(x_1,x_2) = a_2x_1 + b_2x_2 + c_2, \tag{9.8}$$

where A_1 through C_2 are linguistic values and a_1 through c_2 are real numbers. Denoting firing strengths (weights) of these rules by w_1 and w_2, the crisp value, $y_{(C)}$, of the aggregated output signal of the controller is calculated as

$$y_{(C)} = \frac{w_1 f_1(x_1,x_2) + w_2 f_2(x_1,x_2)}{w_1 + w_2} = w_1^{(n)}f_1(x_1,x_2) + w_2^{(n)}f_2(x_1,x_2), \tag{9.9}$$

where $w_1^{(n)} = w_1/(w_1 + w_2)$ and $w_2^{(n)} = w_2/(w_1 + w_2)$ are *normalized weights*.

The neurofuzzy ANFIS controller that realizes Eqs. (9.7) to (9.9) is shown in Figure 9.11. Its architecture consists of five layers, with four

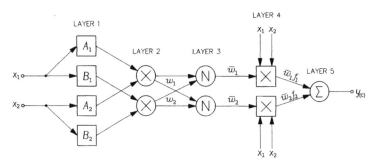

FIGURE 9.11 Neuro-fuzzy ANFIS controller.

nodes in Layer 1; two nodes in each of the Layers 2, 3, and 4; and one node in Layer 5. Every node in Layer 1 is an adaptive node whose output yields the membership grade of the input. The membership functions are parameterized, that is, expressed as functions of the input and certain parameters. This allows the membership functions to be shaped in the tuning process by adjusting the parameters. Nodes in Layer 2 are multipliers, generating the firing strengths, w_1 and w_2, which are next normalized in Layer 3 and used in Layers 4 and 5 to generate the crisp output signal, $y_{(C)}$. Nodes in Layer 4 are adaptive too, that is, parameters a_1 through c_2 in Eqs. (9.8) and (9.9) are adjusted while tuning the controller.

EXAMPLE 9.2   Figure 9.12 shows application of the neurofuzzy controller in an ASD with the space-vector direct torque and flux control (see Section 8.4). The controller produces the reference vector, v_s^*, of stator voltage, expressed in the polar form, $v_s^* = V^* e^{j\alpha^*}$, to be realized in the inverter by a space-vector pulse width modulator. The flux and torque errors, $\Delta\lambda_s$ and ΔT_M, which can be "negative," N, "zero," Z, or "positive," P, and whose membership functions are shown in Figure 9.13, are applied to the controller. A block diagram of the controller is depicted in Figure 9.14.

The reference voltage vector, v_s^*, is obtained as a sum of partial vectors whose magnitudes are generated in the first three layers of the controller. In the fourth layer, an angle is assigned to each nonzero partial vector according to the formula

$$\alpha_i = \Theta_s + \Delta\Theta_i,$$

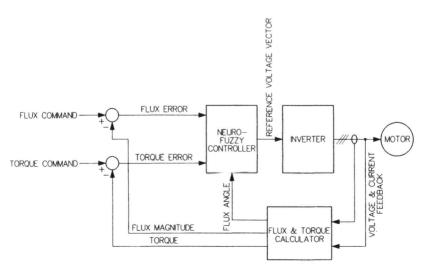

FIGURE 9.12   Block diagram of an SVDTC drive with neuro-fuzzy controller.

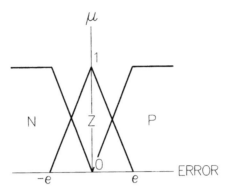

FIGURE 9.13   Membership functions of the flux and torque errors in Example 9.2.

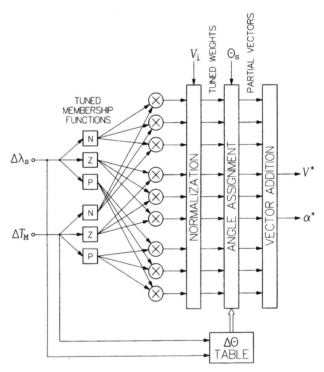

FIGURE 9.14   Neuro-fuzzy controller in Example 9.2.

where α_i and $\Delta\Theta_i$ denote the angle and angle increment for the ith partial vector, respectively, and Θ_s is the angle of stator flux vector. The angle increments are determined according to Table 9.1. Finally, the partial vectors are added in the fifth layer, and the values of V^* and α^* are specified. In the tuning process, the membership functions

TABLE 9.1 Angle Increments in Example 9.2

$\Delta\lambda_s$	P			Z			N		
ΔT_M	P	Z	N	P	Z	N	P	Z	N
$\Delta\Theta$	$\pi/3$	0	$-\pi/3$	$\pi/2$	$\pi/2$	$-\pi/2$	$2\pi/3$	π	$2\pi/3$

in the first layer and weights between the third and fourth layers are adjusted to optimize the operation of the drive system. ∎

9.4 POSITION CONTROL

Position control is the most demanding application of ASDs, requiring an unusually high level of operating performance. The motor starts, runs, and stops, turning by a specific total angle, Θ_M^*. Positioning drives with induction motors are used in elevators and other lifts, material handling equipment, packaging systems, and processing lines. An angular position sensor, usually an encoder, provides the control feedback. The position control loop is the main, outer loop, with the speed and torque loops as inner ones. A general block diagram of a positioning ASD is shown in Figure 9.15. Clearly, the system constitutes an expansion of the speed-controlled drive in Figure 9.2.

Position sensors employed for the control feedback also allow the high-performance indirect field orientation (see Section 7.4). The position controller can be linear (PI or PID), fuzzy or neurofuzzy, or of the variable structure type. A variable structure controller for control of both the speed and position of the motor is shown in Figure 9.16. It is based on the so-called switched-gains concept. The reference torque, T_M^*, is calculated as

$$T_M^* = G_1\Delta\Theta_M + G_2\Delta\omega_M, \qquad (9.10)$$

where $\Delta\Theta_M$ and $\Delta\omega_M$ denote position and speed errors, respectively. Each of the gains G_1 and G_2 is switched between two values, G_1^+, G_1^- and G_2^+, G_2^-, which are either constant or dependent on the state of the drive. Selection of a specific gain depends on the so-called *switching function*, s, given by

$$s = k_1\Delta\Theta_M + k_2\Delta\omega_M, \qquad (9.11)$$

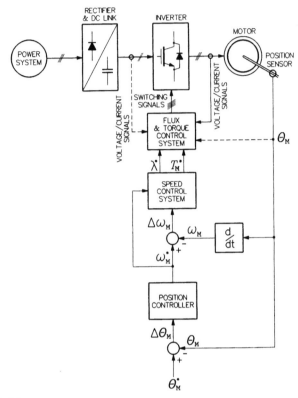

FIGURE 9.15   General block diagram of a high-performance positioning drive.

where coefficients k_1 and k_2 are, again, constant or state dependent. The control law is

$$G_1 = \begin{cases} G_1^+ & \text{if} \quad s\Delta\Theta_M \geq 0 \\ G_1^- & \text{if} \quad s\Delta\Theta < 0 \end{cases}$$

$$G_2 = \begin{cases} G_2^+ & \text{if} \quad s\Delta\omega_M \geq 0 \\ G_2^- & \text{if} \quad s\Delta\omega_M < 0 \end{cases}$$

(9.12)

It can be seen that, in contrast to the system in Figure 9.15, the speed control loop is not subordinated to the position control loop, but both loops are equal in the control hierarchy. This type of speed and position control is typical for certain drives, such as those of elevators, in which the speed profile (speed versus time curve) is so shaped as to ensure a comfortable ride. Energy-optimal positioning drives are another example:

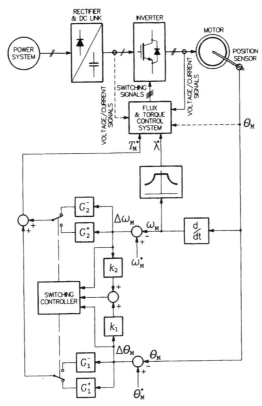

FIGURE 9.16   Positioning ASD with a variable structure speed and position controller.

A parabolic or, if infeasible, a trapezoidal speed profile results in minimum energy consumption during a position change.

9.5 SUMMARY

Most ASDs with induction motors are spindle (controlled-speed) or servo (controlled-position) drives, although systems with autonomous torque control can also be encountered, for instance in winders. Typically, the reference values for the torque and flux control are generated in the speed and position control loops. Under light load conditions, flux control can be used to minimize losses in the motor. The flux must also be adjusted if the drive operates with a speed higher than rated, requiring field weakening.

Various controllers are used for the speed and position control systems, such as linear, variable structure, and machine intelligence controllers.

Variable structure controllers are robust, but the torque command tends to continuously oscillate between two extremes, causing chattering of the speed of the drive. Machine intelligence controllers, of the neural, fuzzy, or neurofuzzy type, emulate human operators and possess precious adaptive properties.

Positioning drives, the most advanced ASDs, usually have three control loops, from the inner, hierarchically lowest, torque control loop, through the speed control loop, to the outer, position control loop. In certain applications, such as the elevator or energy-optimal drives, the speed and position are controlled using parallel loops.

10

SENSORLESS DRIVES

To begin Chapter 10, basic issues in sensorless control of induction motors are outlined. We then present example flux and speed estimators and observers and describe parameter adaptation and self-commissioning procedures. Finally, commercial ASDs with induction motors are reviewed.

10.1 ISSUES IN SENSORLESS CONTROL OF INDUCTION MOTORS

A sensor of angular position of the rotor appears in several block diagrams of induction motor drives described in this book. Knowledge of this position is needed in positioning drives and drives with indirect field orientation. The same sensor can be employed to provide control feedback in speed-controlled drives, the speed signal being usually obtained by differentiating the position signal. Direct speed sensors, such as tachogenerators, are also used in these systems. Some motors in high-performance ASDs are equipped with a built-in encoder hidden inside the motor frame. Motors for drives with direct field orientation may have flux sensors embedded in the stator iron by the air-gap, allowing precise determination of the air-gap flux vector and, consequently, that of the rotor flux vector.

The position, speed, and flux sensors enhance operating characteristics of induction motor drives, but they also increase the cost of the drive and spoil the inherent ruggedness of the induction motor. Therefore, the so-called *sensorless drives* have been receiving a lot of attention. The adjective *sensorless* is actually a misnomer, because no ASD can accurately be controlled without *any* sensors. Because flux sensors are rare in practice, a more suitable name would be "encoderless drives." As it is, the "sensorless" drives are still equipped with voltage and current sensors, signals from which are used in control algorithms. These sensors are inexpensive and installed away from the motor, usually within the power electronic circuitry. As already illustrated in several diagrams (see, for instance, Figure 8.4), typically two current sensors at the inverter output are employed, while a single voltage sensor measures the dc-link voltage, V_i. Individual phase voltages can be reconstructed from V_i and switching variables, a, b, and c, of the inverter using Eqs. 4.3 and 4.8. The same switching variables and a single current sensor in the dc link can be used to reconstruct individual phase currents, but this solution is difficult and uncommon in practice.

Although speed sensors are seldom employed in scalar-controlled drives (position control is beyond capabilities of these systems), the term *sensorless* is customarily limited to vector-controlled ASDs. Manufacturers of induction motor drives classify them, somewhat imprecisely, into three categories: (a) Constant Volts per Hertz (or just Volts per Hertz) or Variable Frequency drives; (b) Sensorless Vector drives; and (c) Field-Oriented drives. Drive systems in the first category are scalar controlled, while the last category offers the highest level of performance. The latter drives, mostly with indirect field orientation, are invariably equipped with speed or position sensors. Although use of the motor itself as a position sensor has been studied, propositions of introducing constructional saliences to the rotor cage still seem to be far from implementation in the manufacturing practice. Speed estimators and observers must rely on the knowledge of motor parameters, being thus inadequate for accurate position estimation. This explains why not all ASDs can be made sensorless.

Generally, sensorless drives are used in speed and torque control applications with moderate performance requirements. The principle of direct-field orientation can be employed if a sufficiently accurate estimation of one of the flux vectors is available. Still, speed ranges of sensorless drives are significantly narrower than those of drives with sensors, because of the difficulties in sensorless control at low speeds. On the other hand, most practical applications of ASDs, such as the ubiquitous pumps, fans, blowers, and compressors, do not really require the highest quality of operation.

Apart from scalar-controlled drives with slip compensation (see Figure 5.6), in which the steady-state speed of the motor can quite precisely be controlled by adjusting the supply frequency, sensorless drives employ a variety of estimators and observers of motor speed, torque, and fluxes. An estimator calculates a given variable in a feedforward manner, that is, using appropriate motor equations into which the measured values of stator voltage and current are substituted. Observers are more sophisticated, with certain self-adjustment features. They are usually based on two or three independent estimators, whose output signals are compared. Their difference, analogous to the control error in closed-loop control systems, is used to adjust signals in the observer until the error is minimized. Examples of flux and speed estimators and observers (termed here jointly as "calculators") used in ASDs with induction motors are given in the subsequent two sections. Note that once a flux vector is identified, the developed torque can easily be calculated from Eqs. (6.17) or (6.18).

10.2 FLUX CALCULATORS

Stator flux vector, λ_s, is given by Eq. (6.15) as a time integral of stator EMF, e_s, given by

$$e_s = v_s - R_s i_s = e_{ds} + je_{qs},\qquad(10.1)$$

and, once known, it allows determination of the rotor flux vector, λ_r, as

$$\lambda_r = \frac{L_r}{L_m}(\lambda_s - \sigma L_s i_s).\qquad(10.2)$$

Estimation of λ_s based on integration of e_s is simple in theory only. The dc drift, caused by noise in the electronic circuits employed, and the difficulty in setting correct initial values of the flux vector components make pure integration impractical. A clever approach to alleviate these problems is based on rearranging Eq. (6.15) from

$$\lambda_s = \frac{1}{p}e_s\qquad(10.3)$$

to

$$\lambda_s = \frac{1}{p + \omega_c}e_s + \frac{\omega_c}{p + \omega_c}\lambda_s,\qquad(10.4)$$

where ω_c is an arbitrary constant. Formally, the first term on the right-hand side of Eq. (10.4) describes a low-pass filter with the cut-off frequency ω_c, which can be adapted to approximate the integrator in practical applications. The second term, subsequently denoted by λ_s'', can be thought of as a feedback signal for compensation of the error introduced by the low-pass filter as a nonideal integrator. The idea of modified algorithms for stator flux estimation is based on the notion that if λ_s'' were properly adjusted, the modified flux estimator could achieve a much better performance than that of the low-pass filter, and the problems associated with pure integration could be avoided.

The simplest solution consists in introducing a saturation block to the feedback loop, as shown in Figure 10.1. For best results, the saturation level should be set to the reference value of the stator flux. Then, the estimator reaches the steady state as soon as after a half-cycle of the flux waveform, even with incorrect initial conditions. Further improvement is obtained by estimating the flux in a revolving reference frame aligned with the flux vector. The block diagram of this estimator is shown in Figure 10.2. The phase angle, Θ_s, of the flux vector required for the dq→DQ transformation [see Eq. (6.21)] is calculated as an integral of the angular speed, ω_e, of that vector. This speed is, in turn, calculated as a ratio of the the Q component, e_{QS}, of the stator EMF to the D component, λ_{DS}, of the stator flux vector. Indeed, if

$$e_s = e_s e^{j\omega_e t} = e_s \cos(\omega_e t) + je_s \sin(\omega_e t) = e_{ds} + je_{qs}, \quad (10.5)$$

then

$$\lambda_s = \frac{e_s e^{j\omega_e t}}{j\omega_e} = \frac{e_s}{\omega_e}\sin(\omega_e t) - j\frac{e_s}{\omega_e}\cos(\omega_e t) = \lambda_{ds} + j\lambda_{qs}, \quad (10.6)$$

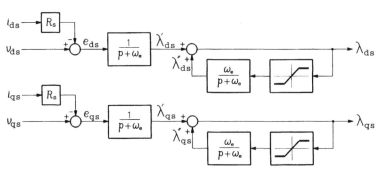

FIGURE 10.1   Stator flux estimator with a saturation block in the feedback loop.

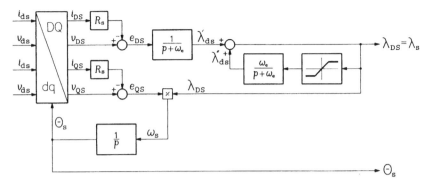

FIGURE 10.2   Stator flux estimator in a revolving reference frame.

and, because a ratio of two vectors does not depend on the reference frame,

$$\frac{e_{QS}}{\lambda_{DS}} = \frac{e_{qs}}{\lambda_{ds}} = \frac{e_s\sin(\omega_e t)}{\frac{e_s}{\omega_e}\sin(\omega_e t)} = \omega_e. \tag{10.7}$$

Another modification of the basic scheme in Figure 10.1 is shown in Figure 10.3, in which the R/P and P/R blocks represent the rectangular to polar and polar to rectangular transformations of the stator flux vector, respectively. In this observer, the estimation error signal, ε, is derived from the condition of orthogonality of space vectors of stator EMF and flux. If the stator EMF vector, e_s, is given by Eq. (10.5), then, from Eq. (10.6),

$$\lambda_{ds} = \lambda_s\sin(\omega_e t) \tag{10.8}$$

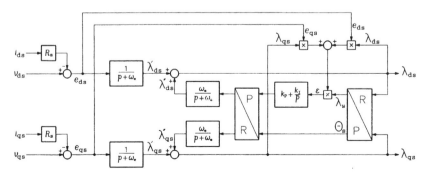

FIGURE 10.3   Stator flux estimator based on the property of orthogonality of the stator EMF and flux vectors.

and

$$\lambda_{qs} = -\lambda_s\cos(\omega_e t). \tag{10.9}$$

Consequently,

$$e_{ds}\lambda_{ds} + e_{qs}\lambda_{qs} = e_s\lambda_s\cos(\omega_e t)\sin(\omega_e t) \tag{10.10}$$
$$- e_s\lambda_s\sin(\omega_e t)\cos(\omega_e t) = 0.$$

In the observer in question, the error signal is defined as

$$\varepsilon = \frac{e_{ds}\lambda_{ds} + e_{qs}\lambda_{qs}}{\lambda_s}, \tag{10.11}$$

and it is applied to a proportional-integral (PI) block to generate the feedback component, λ_s'', of the stator flux vector.

EXAMPLE 10.1 To illustrate the operation of the stator flux observer in Figure 10.3, the example induction motor is considered at $t = 0$, under rated operating conditions. Assuming the phasor of stator voltage, $\hat{V}_s$, to be aligned with the real axis, that is, $\hat{V}_s = 230\angle 0°$ V, the space vector, v_s, of this voltage is $\sqrt{2} \times 1.5 \times 230\angle 0° = 487.9\angle 0°$ V. It means that $v_{ds} = 487.9$ V and $v_{qs} = 0$ V. In Example 6.1, components of the stator current vector under the same conditions were found as $i_{ds} = 75.0$ A and $i_{qs} = -37.4$ A. Thus, components of space vector of the stator EMF are $e_{ds} = 487.9 - 0.294 \times 75.0 = 465.9$ V and $e_{qs} = 0 - 0.294 \times (-37.4) = 11.0$ V, and the EMF vector is $466.0\angle 1.4°$ V. According to Eq. (10.6), the stator flux vector, λ_s, can be determined as $466.0\angle 1.4°/(j377) = 1.236\angle -88.6° = 0.029 - j1.236$ Wb (note that a similar value was already obtained in Example 6.2). If the flux estimation is accurate, the observer error, ε, given by Eq. (10.11), is $[465.9 \times 0.029 + 11.0 \times (-1.236)]/1.236 \approx 0$.

Assuming that the stator flux is overestimated in the observer, for example, $\lambda_s = 1.25\angle -89° = 0.022 - j1.250$ Wb, the observer error is $[465.9 \times 0.022 + 11.0 \times (-1.250)]/1.250 = -2.8$ V. The negative value of ε will cause λ_s'' to decrease, reducing the estimated magnitude of stator flux. Ensuing changes in λ_{ds} and ε will ultimately result in correct values of λ_s and Θ_s. ∎

An interesting technique used in drives with direct field orientation utilizes the phenomenon of magnetic saturation, usually neglected in the idealized model of the induction motor. The saturation of the motor causes induction of a third-harmonic zero-sequence voltage in stator windings.

This component, easily found by adding voltages in individual phases of the stator, allows accurate determination of the spatial position (angle) of the air-gap flux vector.

10.3 SPEED CALCULATORS

Calculation of the motor speed can be performed by a speed estimator using the data of the rotor flux vector, $\lambda_r = \lambda_r \angle \Theta_r$, obtained in a flux calculator. In practice, it is usually the angular velocity, ω_o, of the equivalent two-pole motor that is estimated for the control purposes. Based on rearrangements of dynamic equations of the induction motor, ω_o can be expressed as

$$\omega_o = \frac{d\Theta_r}{dt} - 3\frac{R_r T_M}{P_p \lambda_r^2},$$ (10.12)

where the developed torque, T_M, can be determined from Eq. (6.17) or (6.18). Using the formula for a derivative of the inverse tangent ($\Theta_r = \tan^{-1}(\lambda_{qr}/\lambda_{dr})$), the angular velocity of the rotor flux vector is calculated as

$$\frac{d\Theta_r}{dt} = \frac{1}{\lambda_r^2}\left(\lambda_{dr}\frac{d\lambda_{qr}}{dt} - \lambda_{qr}\frac{d\lambda_{dr}}{dt}\right).$$ (10.13)

Unfortunately, the rotor resistance, R_r, appearing in Eq. (10.12), is the most varying parameter of the motor, due to the wide range of changes in the rotor temperature. Therefore, speed estimators are used in low- and medium-performance ASDs only.

Speed observers, more accurate than speed estimators, are often based on the Model Reference Adaptive System (MRAS) technique. The basic idea of such an observer is illustrated in Figure 10.4. Two estimators

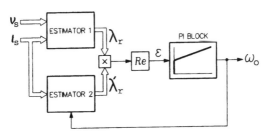

FIGURE 10.4 Configuration of the MRAS speed observer with a PI block.

independently evaluate the rotor flux vector. Estimator 1 uses the so-called *stator equation*, or *voltage equation*,

$$
p\begin{bmatrix} \lambda_{dr} \\ \lambda_{qr} \end{bmatrix} = \frac{L_r}{L_m}\left(\begin{bmatrix} v_{ds} \\ v_{qs} \end{bmatrix}\right.
$$

$$
\left. - \begin{bmatrix} R_s + \sigma L_s p & 0 \\ 0 & R_s + \sigma L_s p \end{bmatrix}\begin{bmatrix} i_{ds} \\ i_{qs} \end{bmatrix}\right)
$$

(10.14)

while the *rotor equation*, or *current equation*

$$
p\begin{bmatrix} \lambda_{dr} \\ \lambda_{qr} \end{bmatrix} = \frac{1}{\tau_r}\left(L_m\begin{bmatrix} i_{ds} \\ i_{qs} \end{bmatrix} - \begin{bmatrix} 1 & \tau_r\omega_o \\ -\tau_r\omega_o & 1 \end{bmatrix}\begin{bmatrix} \lambda_{dr} \\ \lambda_{qr} \end{bmatrix}\right)
$$

(10.15)

is employed in Estimator 2 [Eqs. (10.14) and (10.15) have been derived from Eqs. (6.13), (6.14), and (6.16)]. Because the stator equation does not include the estimated quantity, ω_o, Estimator 1 can be thought of as a reference model and Estimator 2 as an adjustable model. State difference, ε, of the two models is applied to a proportional-integral (PI) block, which produces the estimate of ω_o for the adjustable model. The state difference, that is, the observer error, is given by

$$
\varepsilon = Re\{\boldsymbol{\lambda}_r\boldsymbol{\lambda}_r'\} = \lambda_{qr}\lambda_{dr}' - \lambda_{dr}\lambda_{qr}',
$$

(10.16)

where λ_{dr} and λ_{qr} are estimates of components of the rotor flux vector obtained from Estimator 1, while λ_{dr}' and λ_{qr}' are these generated in Estimator 2. A detailed block diagram of the observer is shown in Figure 10.5.

A neural-network variant of the observer described is shown in Figure 10.6. Here, instead of Estimator 2, a neural model of Eq. (10.15) is employed, in which ω_o constitutes an adjustable weight. The PI block in Figure 10.4 is replaced with a weight tuning algorithm for the neural network.

EXAMPLE 10.2 Operation of the speed observer in Figures 10.4 and 10.5 is illustrated under rated operating conditions of the example motor, at the instant $t = 0$. As seen in Table 2.1, the rated speed of the motor is 1168 r/min, which translates into $\omega_o = 366.9$ rad/. It is assumed that at the considered instant, ω_o is underestimated by 6.9 rad/, that is, Estimator 2 receives the ω_o signal of 360 r/min.

Estimator 1 can be described by the vector equation

$$
\frac{L_m}{L_r}\boldsymbol{\lambda}_r = \frac{1}{p}(v_s - R_s i_s) - \sigma L_s i_s.
$$

(10.17)

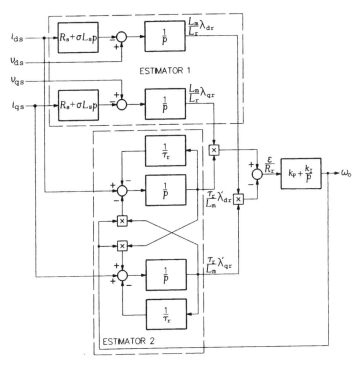

FIGURE 10.5   Block diagram of the MRAS speed observer with a PI block.

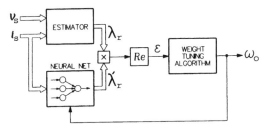

FIGURE 10.6   Configuration of the MRAS speed observer with a neural network.

Because $v_s(0) = 487.9\angle 0°$ V (see Example 10.1), then, with the supply frequency of 60 Hz (377 rad/s), $v_s(t) = 487.9e^{j377t}$ and

$$\frac{1}{p}v_s = \int v_s(t)dt = \int 487.9e^{j377t}dt = \frac{1}{j377} \times 487.9e^{j377t} = \frac{v_s}{j377},$$

that is, the differentiation operator, p, can be replaced by $j377$. Applying this rule to the current vector, $i_s = 83.8\angle -26.5°$ A (see Example 6.1), and completing calculations in Eq. (10.17) yields $(L_m/L_r)\lambda_{dr} = -0.127$ Wb and $(L_m/L_r)\lambda_{qr} = -1.158$ Wb.

Because of the cross-coupling in Estimator 2, calculation of output signals of that estimator must be done separately for the d and q channels. The matrix equation (10.15), can be rewritten as

$$\frac{T_r}{L_m}\lambda'_{dr} = \frac{1}{p}\left(i_{ds} - \frac{1}{L_m}\lambda'_{dr} - \frac{T_r\omega_o}{L_m}\lambda'_{qr}\right)$$

$$\frac{T_r}{L_m}\lambda'_{qr} = \frac{1}{p}\left(i_{qs} - \frac{1}{L_m}\lambda'_{qr} + \frac{T_r\omega_o}{L_m}\lambda'_{dr}\right). \tag{10.18}$$

To evaluate, for instance, i_{ds}/p, the current vector i_s is divided by $j377$, and the real part of the result is extracted. Because $i_s = 75.0 - j37.4$ A, then

$$\frac{i_{ds}}{p} = Re\left\{\frac{i_{ds} + ji_{qs}}{j377}\right\}$$

$$= \frac{1}{377}Re\{i_{qs} - ji_{ds}\}$$

$$= \frac{i_{qs}}{377}$$

$$= \frac{-37.4}{377}$$

$$= -0.0992 \ A/s.$$

Analogously, calculation of an integral of the q component of a vector involves dividing the vector by $j377$ and taking the imaginary part. Thus,

$$\frac{i_{qs}}{p} = Im\left\{\frac{i_{ds} + ji_{qs}}{j377}\right\}$$

$$= \frac{1}{377}Im\{i_{qs} - ji_{ds}\}$$

$$= \frac{-i_{ds}}{377}$$

$$= \frac{-75.0}{377}$$

$$= -0.1989 \ A/s.$$

Applying the rules of integration described above to Eq. (10.18), with $\omega_o = 360$ rad/s, yields

$$\begin{bmatrix} 0.2936 & 0.0647 \\ -0.0647 & 0.2936 \end{bmatrix}\begin{bmatrix} \lambda'_{dr} \\ \lambda'_{qr} \end{bmatrix} = \begin{bmatrix} -0.0992 \\ -0.1989 \end{bmatrix},$$

from which $\lambda'_{dr} = -0.180$ Wb and $\lambda'_{qr} = -0.717$ Wb; that is, $(\tau_r/L_m) \lambda'_{dr} = -1.172$ Wb/Ω and $(\tau_r/L_m) \lambda'_{qr} = -4.669$ Wb/Ω. Consequently,

$$\frac{\varepsilon}{R_r} = (-0.158) \times (-1.172) - (-0.127) \times (-4.669)$$

$$= 0.7642 \text{ Wb}^2/\Omega,$$

and when this signal is applied to the PI block, it will cause the underestimated speed to increase toward the true value. ■

A more advanced adaptive speed observer, based on the theory of *Luenberger state observers*, uses state equations of the induction motor. The state equations are

$$\dot{x} = Ax + Bv_s \tag{10.19}$$

and

$$i_s = Cx,$$

where $x = [i_s \ \lambda_r]^T$,

$$A = \begin{bmatrix} A_{11} & A_{12} \\ A_{21} & A_{22} \end{bmatrix}, \tag{10.20}$$

$B = [B_1 \ 0]^T$, and $C = [I \ 0]$. Elements of matrices A, B, and C are

$$A_{11} = -\frac{1}{\sigma}\left(\frac{1}{\tau_s} + \frac{1-\sigma}{\tau_r}\right)I = a_{r11}I, \tag{10.21}$$

$$A_{12} = \frac{L_m}{\sigma L_s L_r}\left(\frac{1}{\tau_r}I - \omega_o J\right) = a_{r12}I + a_{i12}J, \tag{10.22}$$

$$A_{21} = \frac{L_m}{\tau_r}I = a_{r21}I, \tag{10.23}$$

$$A_{22} = -\frac{1}{\tau_r}I + \omega_o J = a_{r22}I + a_{i22}J, \tag{10.24}$$

and

$$B_1 = \frac{1}{\sigma L_s}I = b_1 I, \tag{10.25}$$

where $\tau_s = L_s/R_s$, $\tau_r = L_r/R_r$,

$$I = \begin{bmatrix} 1 & 0 \\ 0 & 1 \end{bmatrix}, \tag{10.26}$$

and

$$J = \begin{bmatrix} 0 & -1 \\ 1 & 0 \end{bmatrix}. \tag{10.27}$$

The state observer, which simultaneously estimates the stator current and rotor flux, is described by

$$\dot{x}' = A'x' + Bv_s + G(i'_s - i_s), \tag{10.28}$$

where $'$ denotes estimated quantities and G is the gain matrix so selected that the solution of Eq. (10.28) is stable. Specifically,

$$G = \begin{bmatrix} g_1 & -g_2 \\ g_2 & g_1 \\ g_3 & -g_4 \\ g_4 & g_3 \end{bmatrix}, \tag{10.29}$$

where $g_1 = (k - 1)(a_{r11} + a_{r22})$, $g_2 = (k - 1)a_{i22}$, $g_3 = (k^2 - 1)$ $(ca_{r11} + a_{r21}) - c(k - 1)(a_{r11} + a_{r22})$, $g_4 = c(1 - k)a_{i22}$, and $c = \sigma L_s L_r/L_m$. The gains g_1 through g_4 make the observer poles proportional to the motor poles, with k being the coefficient of proportionality. Because the motor is stable, the observer is stable too.

The proportional-integral (PI) adaptive scheme for speed estimation is given by

$$\omega'_o = k_P(e_{ids}\lambda'_{qr} - e_{iqs}\lambda'_{dr}) + k_I \int_0^t (e_{ids}\lambda'_{qr} - e_{iqs}\lambda'_{dr})dt, \tag{10.30}$$

where $e_{ids} = i_{ds} - i'_{ds}$, $e_{iqs} = i_{qs} - i'_{qs}$, and k_P and k_I are the proportional and integral gains, respectively. This scheme improves dynamics of the observer, that is, it allows tracking the actual speed of the motor even when it changes rapidly. The observer is illustrated in Figure 10.7.

Optimal observation of systems in which both the input and output signals are corrupted by noise is provided by *Kalman filters*. It is assumed that the input (measurement) noise and output (process disturbance) noise are uncorrelated and that the system model, here that of the induction

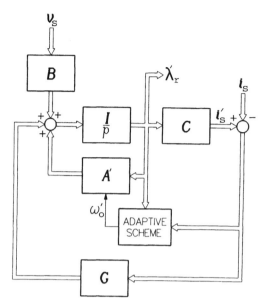

FIGURE 10.7   Configuration of an adaptive speed observer based on state equations of the induction motor.

motor, is accurate. The estimated state is the minimum-variance estimate of the actual state. In the discrete form, for the kth and $k + 1^{\text{th}}$ sampling cycles, motor equations can be written as

$$x(k + 1) = A(k)x(k) + B(k)v(k) + G(k)u(k) \qquad (10.31)$$

and

$$y(k) = C(k)x(k) + w(k),$$

where x, v, and y are state variables (vector components of the stator and rotor currents or stator current and rotor flux), input variables (vector components of the stator voltage), and output variables (vector components of the stator current), respectively. The input noise and output noise vectors are denoted by u and w.

Kalman filter theory applies to linear systems only. Linearization of the employed model of induction machine is based on the assumption that the sampling cycle is so short that the rotor speed (of the equivalent two-pole motor), ω_0, can be considered constant within this cycle. In the Kalman filter algorithm referred to as the Extended Kalman Filter (EKF) and designated for nonlinear systems, the speed is considered as both a parameter and an additional state variable. The EKF equation for estimation of so extended state of the motor is

$$\begin{bmatrix} x(k+1) \\ \omega_o(k+1) \end{bmatrix} = A_e(k)\begin{bmatrix} x(k) \\ \omega_o(k) \end{bmatrix} + B_e(k)v(k) \tag{10.32}$$
$$+ K(k)[y(k+1) - C(k)x(k)],$$

where

$$A_e(k) = \begin{bmatrix} A(k) & 0 \\ 0 & 1 \end{bmatrix} \tag{10.33}$$

and

$$B_e(k) = \begin{bmatrix} B(k) \\ 0 \end{bmatrix}. \tag{10.34}$$

The gain matrix, $K(k)$, is computed at each step in dependence on the noise covariance matrices. For a detailed example of an EKF used in a sensorless ASD, see the paper by Y. R. Kim *et al.* (1994) in the Literature section of this book.

To illustrate the use of speed observers in sensorless ASDs, a speed-control drive with the EKF-based observer is shown in Figure 10.8. Notice the simplicity of the system, in which the speed controller directly

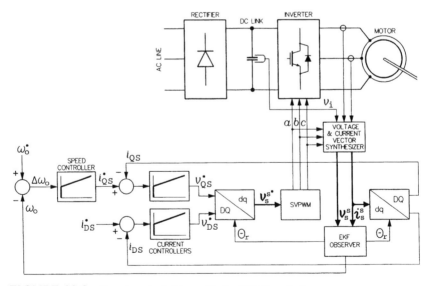

FIGURE 10.8 Speed-control drive with the EKF-based observer.

generates the reference torque-producing current, i_{QS}^*, while the reference flux-producing current, i_{DS}^*, is set to a value corresponding to the rated flux. The current controller yields components, v_{DS}^* and v_{QS}^*, of the reference voltage vector which, with the reference currents, is expressed in the revolving reference frame aligned with the rotor flux vector, also estimated in the observer. The reference voltage vector after transformation to the stator frame is realized by the space-vector PWM. The actual stator voltage vector, $\mathbf{v}_s$, used in the observer is determined from the dc-link voltage, v_i, and switching variables, a, b, and c, of the inverter.

A direct approach to speed estimation uses the motor itself as a speed sensor. Rotor slots produce air-gap permeance waves that affect the air-gap MMF. The rotor speed can be sensed by decomposing the stator current signal into the fundamental component and the slot harmonic component, appearing at frequencies f (supply frequency) and f_{sh}, respectively. Speed ω_o is then determined as

$$\omega_o = \frac{2\pi}{N_s}(f_{sh} + f) \qquad (10.35)$$

where N_s denotes the number of rotor slots. The Fast Fourier Transform (FFT) is employed to find f and f_{sh}. In an even simpler technique, the FFT is used to directly detect the rotational frequency, f_M, of the rotor. Minor imperfections of the rotor, such as imbalance and eccentricity, generate a harmonic component in the spectrum of stator current at that frequency.

10.4 PARAMETER ADAPTATION AND SELF-COMMISSIONING

In high-performance drives with induction motors, accurate information about motor parameters is crucial. Unfortunately, these parameters are not constant, depending on the temperature, magnetic saturation, and supply and slip frequencies. In motors with indirect field orientation, control errors resulting from inaccurate knowledge of motor parameters spoil the orientation, producing steady-state errors and transient oscillations of the flux and torque. In sensorless drives, the flux, torque, and speed calculators may yield incorrect values of these quantities, causing performance deterioration.

The low- and medium-performance drives are often sold without motors. When a given motor is fitted to the drive system, a procedure called *commissioning* must be carried on first. It involves introducing parameters of the motor to the control algorithm and tuning the controllers. Commissioning by the user is inconvenient, and it is desirable that a drive has a self-commissioning capability. For all these reasons, issues of *self-commissioning* and *parameter adaptation*, that is, updating parameter values under changing operating conditions, have been receiving great attention.

Parameter adaptation schemes, for both self-commissioning and control purposes, can be classified as direct and indirect. Direct methods are based on measurement of the parameters in question, sometimes using auxiliary signals superimposed on the stator current. The indirect approach consists in comparing certain motor variables with these computed from the assumed analytical model of the motor. The differences between the real and theoretical values are then minimized by adjusting the parameter estimates.

The simplest self-commissioning consists in estimation of motor parameters based on the nameplate data, using typical per-unit values of these parameters. Clearly, this approach is burdened with inaccuracies. However, it is a reasonably good tool for initialization of parameter adaptation schemes.

The classic *no-load and blocked-rotor tests* for determination of parameters of the steady-state equivalent circuit of induction motor can be adopted for self-commissioning purposes. Both tests utilize the voltage and current measurements to find the impedance of the per-phase equivalent circuits corresponding to these two extremal operating conditions. In particular, with no load, that is, with no rotor current, the equivalent circuit can be reduced to the stator inductance, $L_s = L_m + L_{ls}$ (see Figure 2.14), in series with a no-load motor resistance, R_{NL}, representing power losses in stator copper and iron. The stator resistance, R_s, can be measured directly using an ohmmeter. It can also be determined from a dc voltage applied to two stator terminals and the resultant dc current. With a blocked rotor, the slip, s, equals 1, and the R_s/s resistance is so low that the magnetizing impedance can be neglected. Thus, the equivalent circuit under this condition can be reduced to a series connection of the stator and rotor resistances, $R_s + R_r$, and leakage reactances, $X_{ls} + X_{lr}$. To separate the resistances from reactances, power measurements are taken.

For self-commissioning, prior to normal operation, an ASD can be run with no load (if feasible) to determine the stator inductance. The inverter can be used as a source of dc voltage for measurement of the stator resistance. A direct-on-line start-up can then be arranged, so that the motor, while accelerating from standstill, runs for a short time with the slip close to unity. This emulates the blocked-rotor condition, allowing estimation of the rotor resistance and leakage inductances.

As seen, for instance, from Eq. (7.13), the rotor time constant, τ_r, plays an important role in the indirect field orientation algorithm. A simple way of direct measurement of this parameter consists in exciting the stator with a dc current, so that the magnetic field builts up in the motor, and then turning the inverter off. The stator is now open circuited, and the field dies out, the magnetic energy being dissipated only in the rotor. A low voltage induced in the stator fades away exponentially with the time constant equal to τ_r.

An iterative, on-line approach to the rotor time constant estimation in an indirect field orientation drive is based on Eq. (7.15). The vector diagram of a field-oriented induction motor is shown in Figure 10.9 (in which e_r denotes the rotor EMF vector). Eq. (7.15) can be rewritten as

$$\tau_r = \frac{1}{\omega_r^*} \frac{i_{DS}}{i_{DS}} = \frac{\tan(\Theta_s)}{\omega_r^*}, \qquad (10.36)$$

where Θ_s denotes the phase angle of stator current vector, i_s, in the revolving reference frame aligned with the rotor flux vector, λ_r. From the vector diagram,

$$\tan(\Theta_s) = \frac{v_s \cos(\varphi) - R_s i_s}{v_s \sin(\varphi) - \omega L_s \sigma i_s}, \qquad (10.37)$$

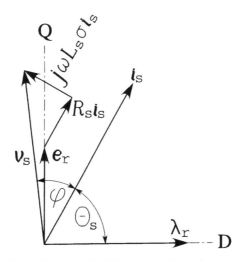

FIGURE 10.9 Vector diagram of a field-oriented induction motor.

where φ is the angle between vectors of stator voltage and stator current. In the described estimator, quantities v_s, i_s, and φ are computed from the reference values of vectors in question, while ω is determined as a sum of the reference rotor speed, ω_o^*, and the reference slip speed, ω_r^*. It can be seen that sequential values of the reference slip speed serve to estimate values of the rotor time constant which, in turn, is used to compute the reference slip speed. Even when started with a highly inaccurate estimate of τ_r, this iterative process quickly converges to the actual value of rotor time constant.

If a rotor speed sensor is used, continuous estimation of motor parameters can be done using the scheme shown in Figure 10.10. Stator voltage and current vectors are stored and, with ω_o, applied to the analytical model of the motor given by Eq. (6.7). The computed vector, i_s', of stator current is compared with the real current vector, i_s, and the resultant adaptation error, ε, is minimized in an iterative process. The values of motor parameters in the analytical model are adjusted using the method of maximum error gradient.

Motor state observers can be used for parameter adaptation by treating the estimated parameter as an additional state variable. Adaptive and EKF-based speed observers are particularly suited for that purpose. For example, the Luenberger-type speed observer in Figure 10.7 can also be used for identification of the stator resistance and rotor time constant. The respective adaptive schemes are given by

$$\frac{d}{dt}R_s' = -k_R(e_{ids}i_{ds}' + e_{iqs}i_{qs}'),$$ (10.38)

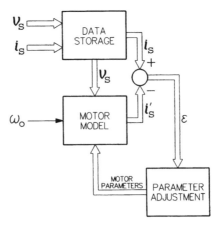

FIGURE 10.10   Motor parameter adaptation scheme when speed of the rotor is known.

and

$$\frac{d}{dt}\left(\frac{1}{\tau_r'}\right) = \frac{k_\tau}{L_r[e_{ids}(\lambda_{dr}' - L_m i_{ds}') + e_{iqs}(\lambda_{qr}' - L_m i_{qs}')]}, \quad (10.39)$$

where k_R and k_τ are arbitrary positive constants (see Section 10.3 for explanation of the other symbols).

Parameter adaptation schemes significantly increase the complexity of control systems. Therefore, various alternative approaches to adaptive tuning of vector control systems have been proposed. For example, a simple technique using a torque feedback to adjust the reference rotor frequency, ω_r^*, signal in drives with indirect field orientation is illustrated in Figure 10.11. As seen from Eq. (7.15), ω_r^* is inversely proportional to the rotor time constant, τ_r, one of the most unstable parameters of the induction machine. In the adaptive scheme described, the ω_r^* signal is corrected by subtracting from it an adaptation signal, $\Delta\omega_r^*$. It is generated by a PI-type controller, whose input signal is the adaptation error, ε_2, obtained by multiplying a control error, ε_1, by the reference torque-producing current, i_{QS}^*. The control error represents the difference between absolute values of the reference torque, T_M^*, and the estimated torque, T_M, of the motor. This scheme improves the torque control by weakening the dependence of ω_r^* on the rotor time constant.

Another idea of robust field-orientation drive uses a high-frequency current, $i_h = I_h \sin(\omega_h t)$, superimposed on the D-axis stator current, i_{DS}, in the excitation frame aligned with the rotor flux vector, $\lambda_r^e = \lambda_{DR} +$

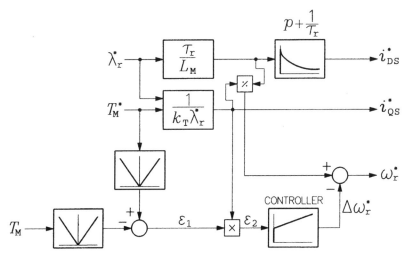

FIGURE 10.11 Adaptive control of the reference rotor frequency.

$j\lambda_{QR}$. The high-frequency current barely affects the rotor flux, thanks to the low-pass action of the magnetizing inductance, but it causes a torque ripple, T_h. From Eq. (6.18),

$$T_h = k_T(i_{QS}\lambda_{DR} - i_{DS}\lambda_{QR}), \tag{10.40}$$

where $k_T = 2p_pL_m/(3L_r)$. Eq. (10.40) can be rewritten as

$$T_h = -k_T\lambda_r\sin(\Delta\Theta_r) \times I_h\sin(\omega_h t), \tag{10.41}$$

where λ_r is the magnitude of rotor flux and $\Delta\Theta_r$ denotes the deviation of rotor flux vector from ideal orientation. Multiplying T_h by i_h yields

$$T_h i_h = -k_T\lambda_r\sin(\Delta\Theta_r) \times I_h^2\sin^2(\omega_h t) \tag{10.42}$$

$$= -k_T\lambda_r\sin(\Delta\Theta_r) \times I_h^2\frac{1 - 2\cos(2\omega_h t)}{2}.$$

When a low-pass filter is employed to eliminate the $\cos(2\omega_h t)$ term, Eq. (10.42) can be replaced with

$$LP\{T_h i_h\} = -\frac{k_T}{2}I_h^2\lambda_r\sin(\Delta\Theta_r), \tag{10.43}$$

where $LP\{...\}$ denotes output of the low-pass filter. Taking into account that $\sin(\Delta\Theta_r) \approx \Delta\Theta_r$,

$$\Delta\Theta_r \approx -\frac{2}{k_T\lambda_r I_h^2}LP\{T_h i_h\}. \tag{10.44}$$

The angular position of the rotor flux vector, Θ_r, required for the field orientation is determined by the digital control system as $\Theta_r(k + 1) = \Theta_r(k) + \Delta\Theta_r$. It can be seen that no information on the rotor time constant is required. However, the ripple torque, T_h, appearing in Eq. (10.44), as well as the magnitude, λ_r, of rotor flux vector, must be estimated precisely. The "pseudointegral" approach described in Section 10.2 [see Eq. (10.4)] can be used to calculate λ_r, as

$$\lambda_r = \left|\frac{L_r}{L_m}\frac{v_s - R_s i_s}{p + \omega_c} - L_\sigma i_s\right|, \tag{10.45}$$

where $L_\sigma = L_s - L_m^2/L_r \approx L_{ls} + L_{lr}$. Then, T_h can be determined as

$$T_h = HP\{k_t(i_{qs}\lambda_{dr} - i_{ds}\lambda_{qr})\}, \tag{10.46}$$

where $HP\{...\}$ denotes output of a high-pass filter.

The L_r/L_m ratio in Eq. (10.46) does not vary significantly (both inductances change simultaneously under the influence of operating conditions), and the stator resistance can easily be updated when the drive is at standstill. Therefore, only the inductance L_σ must be subjected to parameter adaptation. From the dynamic equivalent circuit of the induction motor,

$$v_h = (R_s + R_r)i_h + \omega_h(L_{ls} + L_{lr})I_h\cos(\omega_h t), \qquad (10.47)$$

where v_h denotes the high-frequency component of the stator voltage producing the high-frequency current, i_h. Eq. (10.47) can be rewritten as

$$[v_h - (R_s + R_r)i_h]^2 = \qquad (10.48)$$
$$[\omega_h(L_{ls} + L_{lr})I_h]^2 \frac{1 + \cos(2\omega_h t)}{2},$$

and, when low-pass filtering is applied, L_σ is obtained as

$$L_\sigma = \sqrt{2LP\left\{\left[\frac{v_h - (R_s + R_r)i_h}{\omega_h i_h}\right]2\right\}}. \qquad (10.49)$$

Accurate estimation of L_σ requires that $\omega_h L_\sigma \gg R_s + R_r$, which is easily satisfied with a sufficiently high frequency, ω_h, of the injected current.

10.5 COMMERCIAL ADJUSTABLE SPEED DRIVES

A book on control of induction motors would be incomplete without a short overview of commercial ASDs available on the market. The majority of these drives are sensorless. If available, the speed or position sensors are usually offered as an option. Most of the drives are used in industry, but domestic and vehicular applications of induction motor ASDs are on the rise.

The growth in controlled induction motor drives over the last decade has been greater than that in the previous three decades. It has been mostly a response to demands for: (a) increased efficiency and reliability of electromechanical power conversion, (b) process automation, and (c) flexibility of the power train. The rapidly decreasing cost of information processing (dedicated DSP controllers for motion control can be purchased for less than $5 apiece) and the continuing improvement in operating characteristics of semiconductor power switches are additional growth-supporting factors. In 1998, worldwide sales of ac ASDs were close to five billion dollars. The highest saturation of industry by induction motor drives occurs in Japan, followed by Europe and the United States. As time progresses, the drives become more compact and less expensive.

Voltage source inverters tend to phase out the current source inverters. The inverters operate mostly in the PWM mode, but in the field-weakening region, where the highest stator voltage is needed, the square-wave operation is employed. IGBTs are the most common semiconductor power switches. Modern IGBTs have rise times of 5 to 10 kV/μs, voltage ratings up to 3.3 kV, and current ratings up to 2.4 kA. Consequently, they are able to replace GTOs in many high-power applications.

Apart from the control capabilities described in this book, commercial ASDs have several other operation-enhancing features. In most drives, the dc-link voltage is sensed and the drive is shut down when the voltage drops below a predetermined level. However, in so-called *critical drives*, measures are applied to maintain the drive operational when, as often happens, the supply power is lost for only a second or so. These include a flywheel, an extra large dc-link capacitance, and an override of the automatic shutdown. When the power reappears, the motor is automatically accelerated to the set speed.

Electric braking is used in most ASDs supplied from the power line via a diode rectifier, incapable of transferring power from the drive to the line. Three basic approaches are: (a) *dynamic braking*, using an external resistor connected in parallel with the dc-link capacitor by an extra switch (see Figure 4.20); (b) *flux braking*, used below the rated frequency and consisting in increasing the flux to convert portion of kinetic energy of the drive into iron losses in the motor; and (c) *dc-current braking*, by injection of a dc current into stator windings.

With the increase in switching frequencies, the dead time (see Section 4.3) in the inverter is no longer negligible, and it can compromise the voltage and current control. When the polarity of output currents is known, the dead time can be compensated by appropriate adjustment of the switching times. Another method consists in comparing the output voltage integral (virtual flux) with a reference integral. This feedback arrangement allows continuous correction of the voltage error.

As already mentioned, drive manufacturers classify induction motor ASDs into three categories: (a) Constant Volts per Hertz (CVH) drives, (b) Sensorless Vector (SV) drives, and (c) Field-Oriented (FO) drives. The CVH ASDs are based on the open-loop voltage and frequency control (see Section 5.2). The current is sensed for protection purposes only, usually in the dc link. Dynamics at low speeds is poor. However, typical applications of CVH drives, such as pumps, blowers, compressors, and fans, do not really need precise speed or torque control.

Many general purpose ASDs have the user-selectable Sensorless Vector option. SV drives, although somewhat inferior to systems with the speed or position feedback, are a good choice for applications requiring decent dynamic performance at low speeds. Specifically, low static errors and overshoots, fast reaction to command changes, and wide ranges of torque and speed control are called for. The SV ASDs are employed in printing lines (1:100 speed range), paper machines (1% control error tolerance), steel mills (1:50 speed range), and coating machinery (smooth speed changes). Typical parameters of commercial sensorless drives are:

- speed range from 1:5 to 1:120
- static speed error at the minimum/rated speeds: -3.0% to -0.1% / 1.6% to 0.0%
- dynamic speed error (at 30-Hz frequency): 2.5%/s to 0.25%/s
- speed control bandwidth (at 30 Hz frequency): 10 rad/s to 20 rad/s
- starting torque from 100% to 150% of the rated torque

For sensorless drives, commissioning procedures are necessary to determine parameters of the motor used in control algorithms and to set controller gains. Performance requirements dictate accuracy of the parameters, extent of the commissioning tests, and feedback resolution. Many manufacturers of ASDs provide complete drive systems, including motors, which facilitates the commissioning. Certain self-commissioning and auto-tuning functions are incorporated in advanced drives. Fine tuning is also available through *user interface* devices, including Windows-based software for interfacing the drive with a personal computer.

The highest level of performance is obtained in Field-Oriented ASDs, also called Flux Vector Control or Full Vector drives, equipped with encoders or resolvers. Speed ranges of FO drives are much wider than these of other ASDs with induction motors, approaching the 1:20,000 value. Bandwidths of the torque and speed control are an order of magnitude greater than those of the sensorless drives; for instance, 1000 rad/s for the torque loop and 100 rad/s for the speed loop. Speed control errors can be as low as 0.01% of the rated speed. Motors produced especially for FO drives, apart from an independent cooling fan, often have an optional integrated encoder. These motors are more expensive than the standard motors, but they can operate in similarly harsh environments. Certain manufacturers of ASDs also equip their motors with a spring-loaded friction brake for rapid stopping of the drive system. High-quality gear boxes of various ratios and configurations are offered with the motors.

The progress in current control techniques and fast semiconductor power switches has made PWM voltage-source inverters to evolve into high-bandwidth power amplifiers. Electric power conversion efficiency at rated power approaches 98%. Selectable switching frequencies are common. Typical range of output frequencies is 0 to 240 Hz, but frequencies as high as 800 Hz are available in certain drives. For multimotor drive systems, single dc sources, usually with the reverse power flow capability, are employed to supply several inverters. Certain small drives have the power electronic and control circuits integrated with the motor.

Typical protection systems include:

- phase-loss protection
- overvoltage protection
- undervoltage tripping (with ride-through options)
- line reactors for protection from the supply voltage transients
- overcurrent protection (plus input fuses)
- short-circuit and ground fault protection
- overtemperature protection
- motor overload (stall) protection

Data recorders and displays allow storing and displaying vital information about normal and faulty operating conditions of the drive. Critical ASDs may be equipped with the Essential Service Override (ESO): In the case of an emergency, the drive is required to run as long as it can, even under the threat of complete destruction.

Automated production lines employ several ASDs, whose operation must be precisely synchronized. Control systems of such drives include communication networks using various communication protocols.

Low-voltage (230 V, 460 V, 575 V) drives are offered in a wide, 0.1 kW to 1.5 MW power range. Rated power of *medium voltage* (2.3 kV to 7.2 kV) ASDs may be as high as 20 MW. Three-level inverters are often used in these drives. As of now, among the major drive manufacturers only ABB offers DTC drives, in the wide range of 2.2 kW to 4.3 MW of rated power.

10.6 SUMMARY

Sensorless drives operate without speed or position sensors, which increases their robustness and reduces the equipment cost. Quality of operation of sensorless drives falls between that of scalar control drives and vector control drives with speed/position sensors.

Estimators and observers of flux, torque, and speed constitute crucial components of control systems for sensorless ASDs. Estimators, based on dynamic equations of the induction motor, operate in the feedforward manner, while observers, which use certain feedback arrangements, are capable of correction of their estimates. Correct information about motor parameters is also necessary for proper operation of sensorless drives. Therefore, parameter adaptation and self-commissioning are important parts of the control process in these drives.

Commercial ASDs can be divided into three major classes, namely, Constant Volts per Hertz, Sensorless Vector, and Field-Oriented drives. In addition to the flux, torque, speed, and position control systems, commercial drives possess numerous features enhancing their operation, and they are equipped with various protection systems. The drives cover a wide range of rated power, from a fraction of one kilowatt to tens of megawatts.

LITERATURE

The amount of technical literature devoted to ac ASDs is very large, most of it having appeared within the last two decades. The majority of the publications cover various aspects of control of induction motors. In addition to dedicated books, the following are the best sources of information in English.

JOURNALS

Electric Machines and Power Systems
European Power Electronics Journal
IEEE Transactions on Industry Applications
IEEE Transactions on Industrial Electronics
IEEE Transactions on Energy Conversion
IEEE Transactions on Power Electronics
IEE Proceedings, part B

CONFERENCE PROCEEDINGS

Annual IEEE Industry Applications Society Meeting (IEEE-IAS)
Annual IEEE Industrial Electronics Society Conference (IECON)
Annual Power Electronics Conference and Exposition (APEC)
European Conference on Power Electronics and Applications (EPE)
International Conference on Electric Machines (ICEM)
International Electric Machines and Drives Conference (IEMDC)
Power Electronics Specialists Conference (PESC)

The Literature listed below has been selected on the basis of availability, readability, technical importance, and tutorial value of the contents. By no means is the bibliography complete, and readers are strongly encouraged to probe further. Nomenclature used in this book is shared by most authors, with minor variations in some symbols (e.g., $\alpha\beta$ coordinates instead of dq coordinates). In many publications, space vectors of motor variables are scaled down by a factor of $1.5\sqrt{2}$, so that the magnitude of a vector equals the rms value of the corresponding phasor. Consequently, in such a case, torque equations must be scaled up by a factor of 4.5.

BOOKS

Boldea, I., and Nasar, S. A., *Vector Control of AC Drives*, CRC Press, Boca Raton, Florida, 1992.

Boldea, I., and Nasar, S. A., *Electric Drives,* CRC Press, Boca Raton, Florida, 1999.

Bose, B. K., *Power Electronics and AC Drives*, Prentice-Hall, Englewood Cliffs, N.J., 1986.

Bose, B. K. (editor), *Microcomputer Control of Power Electronics and Drives*, IEEE Press, New York, 1987.

Bose, B. K. (editor), *Power Electronics and Variable Frequency Drives*, IEEE Press, New York, 1996.

Chapman, S. J., *Electric Machinery Fundamentals,* 3rd ed., McGraw-Hill, Boston, 1999.

Kazmierkowski, M. P., and Tunia, H., *Automatic Control of Converter-Fed Drives*, Elsevier, Amsterdam, 1994.

Krause, P. C., and Wasynczuk, O., *Electromechanical Motion Devices*, McGraw-Hill, New York, 1989.

Leonhard W., *Control of Electrical Drives*, 2nd ed., Springer-Verlag, Berlin, 1996.

Murphy, J. M. D., and Turnbull, F. G., *Power Electronic Control of AC Motors*, Pergamon Press, Oxford, 1988.

Novotny, D. W., and Lipo, T. A., *Vector Control and Dynamics of AC Drives*, Oxford University Press, Oxford, 1996.

Novotny, D. W., and Lorenz, R. D. (editors), *Introduction to Field Orientation and High Performance AC Drives*, Tutorial Course, IEEE Industry Applications Society, 1986.

Ong, C. M., *Dynamic Simulation of Electric Machinery Using Matlab/ Simulink*, Prentice Hall, Englewood Cliffs, N.J. 1997.

Rajashekara, K., Kawamura, A., and Matsuse, K. (editors), *Sensorless Control of AC Motor Drives*, IEEE Press, New York, 1996.

Rosenberg, R., and Hand, A., *Electric Motor Repair*, 3rd ed., Holt, Rinehart and Winston, New York, 1986.

Stefanovic, V. R., and Nelms, R. M. (editors), *Microprocessor Control of Motor Drives and Power Converters*, Tutorial Course, IEEE Industry Applications Society, 1992.

Trzynadlowski, A. M., *The Field Orientation Principle in Control of Induction Motors*, Kluwer Academic Publishers, Boston, 1994.

Trzynadlowski, A. M., *Introduction to Modern Power Electronics*, John Wiley, New York, 1998.

Vas, P., *Vector Control of AC Machines*, Oxford University Press, Oxford, 1990.

Vas, P., *Sensorless Vector and Direct Torque Control*, Oxford University Press, Oxford, 1998.

PAPERS

Aoki, N., Satoh, K., and Nabae, A., "Damping circuit to suppress motor terminal overvoltage and ringing in PWM inverter fed ac motor drive systems with long motor leads," *IEEE Transactions on Industry Applications*, vol. 35, no. 5, pp. 1014–1020, 1999.

Atkinson, D. J., Acarnley, P. P., and Finch, J. W., "Observers for induction motor state and parameter estimation," *IEEE Transactions on Industry Applications*, vol. 27, no. 6, pp. 1119–1127, 1991.

Attaianese, C., Damiano, A., Gatto, G., Marongiu, I., and Perfetto, A., "Induction motor drive parameters identification," *IEEE Transactions on Power Electronics*, vol. 13, no. 6, pp. 1112–1122, 1998.

Attaianese, C., Nardi, V., Perfetto, A., and Tomasso, G., "Vectorial torque control: A novel approach to torque and flux control of induction motor drives," *IEEE Transactions on Industry Applications*, vol. 35, no. 6, pp. 1399–1405, 1999.

Baader, U., Depenbrock, M., and Gierse, G., "Direct self-control (DSC) of inverter-fed induction machine: Basis for speed control without speed-measurement," *IEEE Transactions on Industry Applications*, vol. 28, no. 3, pp. 581–588, 1992.

Ba-Razzouk, A., Cheriti, A., Olivier, G., and Sicard, P., "Field oriented control of induction motors using neural-network decouplers," *IEEE Transactions on Power Electronics*, vol. 12, no. 4, pp. 752–763, 1997.

Ben-Brahim, L., and Kawamura, A., "A fully digitized field-oriented controlled induction motor drive using only current sensors," *IEEE Transactions on Industrial Electronics*, vol. 39, no. 3, pp. 241–249, 1992.

Ben-Brahim, L., Tadakuma, S., and Akdag, A., "Speed control of induction motor without rotational transducers," *IEEE Transactions on Industry Applications*, vol. 35, no. 4, pp. 844–850, 1999.

Blaschke, F., "The principle of field-orientation as applied to the new 'transvektor' closed-loop control system for rotating-field machines," *Siemens Review*, vol. 34, nr. 5, pp. 217–220, 1972.

Blaschke, F., Van der Burgl, I., and Vandenput, A., "Sensorless direct field orientation at zero flux frequency," *Conference Record of IEEE-IAS'96*, pp. 189–196.

Bodson, M., Chiasson, J., and Novotnak, R. T., "Nonlinear speed observer for high-performance induction motor control," *IEEE Transactions on Industrial Electronics*, vol. 42, no. 4, pp. 337–343, 1995.

Boldea, I., and Nasar, S. A., "Torque vector control (TVC)—A class of fast and robust torque-speed and position digital controllers for electric drives," *Electric Machines and Power Systems*, vol. 15, no. 3, pp. 209–223, 1988.

Bonnano, F., Consoli, A., Raciti, A. and Testa, A., "An innovative direct self-control scheme for induction motor drives," *IEEE Transactions on Power Electronics*, vol. 12, no. 5, pp. 800–806, 1997.

Bose, B. K., "Expert system, fuzzy logic, and neural network applications in power electronics and motion control," *Proceedings of the IEEE*, vol. 82, no. 8, pp. 1303–1323, 1994.

Bose, B. K., and Patel, N. R., "A programmable cascaded low-pass filter-based flux synthesis for a stator flux-oriented vector-controlled induction motor drive," *IEEE Transactions on Industrial Electronics*, vol. 44, no. 1, pp. 140–143, 1997.

Bose, B. K., and Patel, N. R., "Quasi-fuzzy estimation of stator resistance of induction motor," *IEEE Transactions on Power Electronics,* vol. 13, no. 3, pp. 401–409, 1998.

Bose, B. K., Patel, N. R., and Rajashekara, K., "A start-up method for a speed sensorless sator-flux-oriented vector-controlled induction motor drive," *IEEE Transactions on Industrial Electronics,* vol. 44, no. 4, pp. 587–589, 1997.

Burton, B., Kamran, F., Harley, R. G., Habetler, T. G., Brooke, M. A., and Poddar, R., "Identification and control of induction motor stator currents using fast on-line random training of a neural network," *IEEE Transactions on Industry Applications,* vol. 33, no. 3, pp. 697–704, 1997.

Busse, D. F., Erdman, J. M., Kerkman, R. J., Schlegel, D. W., and Skibinski, G. L., "The effects of PWM voltage source inverters on the mechanical performance of rolling bearings," *IEEE Transactions on Industry Applications,* vol. 33, no. 2, pp. 567–576, 1997.

Cabrera, L. A., Elbuluk, M. E., and Husain, I., "Tuning the stator resistance of induction motors using artificial neural network," *IEEE Transactions on Power Electronics,* vol. 12, no. 5, pp. 779–787, 1997.

Cabrera, L. A., Elbuluk, M. E., and Zinger, D. S., "Learning techniques to train neural networks as a state selector for inverter-fed induction machines using direct torque control," *IEEE Transactions on Power Electronics,* vol. 12, no. 5, pp. 788–799, 1997.

Cerruto, E., Consoli, A., Raciti, A., and Testa, A., "Fuzzy adaptive vector control of induction motor drives," *IEEE Transactions on Power Electronics,* vol. 12, no. 6, pp. 1028–1040, 1997.

Chang, J. H., and Kim, B. K., "Minimum-time minimum-loss speed control of induction motors under field-oriented control," *IEEE Transactions on Industrial Electronics,* vol. 44, no. 6, pp. 809–815, 1997.

Chiasson, J., "Dynamic feedback linearization of the induction motor," *IEEE Transactions on Automatic Control,* vol. 38, no. 10, pp. 1588–1594, 1993.

Chiasson, J., "New approach to dynamic feedback linearization control of an induction motor," *IEEE Transactions on Automatic Control,* vol. 43, no. 3, pp. 391–397, 1998.

Chrzan, P. J., and Kurzynski, P., "A rotor time constant evaluation for vector-controlled induction motor drives," *IEEE Transactions on Industrial Electronics,* vol. 39, no. 5, pp. 463–465, 1992.

Cilia, J., Asher, G. M., Bradley, K. J., and Sumner, M., "Sensorless position detection for vector controlled induction motor drives using an asymmetric outer-section cage," *IEEE Transactions on Industry Applications*, vol. 33, no. 5, pp. 1162–1169, 1997.

Degner, M. W., and Lorenz, R. D., "Using multiple saliences for the estimation of flux, position, and velocity in ac machines," *IEEE Transactions on Industry Applications*, vol. 34, no. 5, pp. 1097–1104, 1997.

Depenbrock, M., "Direct self control (DSC) of inverter-fed induction machine," *IEEE Transactions on Power Electronics*, vol. 3, no. 4, pp. 420–429, 1988.

Doki, S., Takahashi, K., and Okuma, S., "Slip-frequency type and flux-feedback type vector controls in discrete-time system," *IEEE Transactions on Industrial Electronics*, vol. 44, no. 3, pp. 382–389, 1997.

Epperly, R. A., Hoadley, F. L., and Piefer, R. W., "Considerations when applying ASD's in continuous processes," *IEEE Transactions on Industry Applications*, vol. 33, no. 2, pp. 389–396, 1997.

Erdman, J. M., Kerkman, R. J., Schlegel, D. W., and Skibinski, G. L., "Effect of PWM inverters on ac motor bearing currents and shaft voltages," *IEEE Transactions on Industry Applications*, vol. 32, no. 2, pp. 250–259, 1996.

Espinosa-Perez, G., Campos-Canton, I., and Ortega, R., "On the theoretical robustness of a passivity-based controller for induction motors," *Proceedings of 1996 IEEE International Conference on Control Applications*, pp. 626–631, 1996.

Famouri, P., and Cathey, J. J., "Loss minimization control of an induction motor drive," *IEEE Transactions on Industry Applications*, vol. 27, no. 1, pp. 32–37, 1991.

Ferrah, A., Bradley, K. J., Hogben-Laing, P. J., Woolfson, M. S., Asher, G. M., Sumner, M., and Shuli, J., "A speed identifier for induction motor drives using real-time adaptive digital filtering," *IEEE Transactions on Industry Applications*, vol. 34, no. 1, pp. 156–162, 1997.

Gökdere, L. U., and Simaan, M. A., "A passivity-based method for induction motor control," *IEEE Transactions on Industrial Electronics*, vol. 44, no. 5, pp. 688–695, 1997.

Griva, G., Habetler, T. G., Profumo, F., and Pastorelli, M., "Performance evaluation of a direct torque controlled drive in the continuous PWM-square wave transition range," *IEEE Transactions on Power Electronics*, vol. 10, no. 4, pp. 464–471, 1995.

Grotstollen, H., and Wiesing, J., "Torque capability and control of a saturated induction motor over a wide range of flux weakening," *IEEE Transactions on Industrial Electronics*, vol. 42, no. 4, pp. 374–381, 1995.

Habetler, T. G., and Divan, D. M., "Control strategies for direct torque control using discrete pulse modulation," *IEEE Transactions on Industry Applications*, vol. 27, no. 4, pp. 893–901, 1991.

Habetler, T. G., Profumo, F., Griva, G., Pastorelli, M., and Bettini, A., "Stator resistance tuning in a stator-flux field-oriented drive using an instantaneous hybrid flux estimator," *IEEE Transactions on Power Electronics*, vol. 13, no. 1, pp. 125–133, 1998.

Habetler, T. G., Profumo, F., Pastorelli, M., and Tolbert, L. M., "Direct torque control of induction machines using space vector modulation," *IEEE Transactions on Industry Applications*, vol. 28, no. 5, pp. 1045–1053, 1992.

Harnefors, L., and Nee H. P., "Model-based current control of ac machines using the internal model control method," *IEEE Transactions on Industry Applications*, vol. 34, no. 1, pp. 133–141, 1997.

Hasse, K., "About the dynamics of adjustable-speed drives with converter-fed squirrel-cage induction motors" (in German), Ph.D. Dissertation, *Darmstadt Technische Hochschule*, 1969.

Heber, B., Xu, L., and Tang, Y., "Fuzzy logic enhanced speed control of an indirect field-oriented induction motor drive," *IEEE Transactions on Power Electronics,* vol. 12, no. 5, pp. 772–778, 1997.

Hickok, H. N., "Adjustable speed—A tool for saving energy losses in pumps, fans, blowers, and compressors," *IEEE Transactions on Industry Applications*, vol. 21, no. 1, pp. 124–136, 1985.

Hofmann, H., and Sanders, S. R., "Speed-sensorless vector torque control of induction motors using a two-time scale approach," *IEEE Transactions on Industry Applications*, vol. 34, no. 5, pp. 169–177, 1997.

Hofmann, H., Sanders, S. R., and Sullivan, C. R., "Stator-flux based vector control of induction machines in magnetic saturation," *IEEE Transactions on Industry Applications*, vol. 33, no. 4, pp. 935–942, 1997.

Hofmann, W., and Krause, M., "Fuzzy control of ac-drives fed by PWM-inverters," *Proceedings of IECON'92*, pp. 82–87, 1992.

Holtz, J., "Speed estimation and sensorless control of ac drives," *Proceedings of IECON'93*, pp. 649–654, 1993.

Holtz, J., "The representation of ac machine dynamics by complex signal flow graphs," *IEEE Transactions on Industrial Electronics*, vol. 42, no. 3, pp. 263–271, 1995.

Holtz, J., "On the spatial propagation of transient magnetic fields in a.c. machines," *IEEE Transactions on Industry Applications*, vol. 32, no. 4, pp. 927–937, 1996.

Holtz, J., and Thimm, T., "Identification of machine parameters in a vector-controlled induction motor drive," *IEEE Transactions on Industry Applications*, vol. 27, no. 6, pp. 1111–1118, 1991.

Hsieh, G. C., and Hung, J. C., "Phased-locked loop techniques—A survey," *IEEE Transactions on Industrial Electronics*, vol. 43, no. 6, pp. 609–615, 1996.

Hu, J., and Wu, B., "New integration algorithms for estimating motor flux over a wide speed range," *IEEE Transactions on Power Electronics*, vol. 13, no. 5, pp. 969–977, 1998.

Hurst, K. D., and Habetler, T. G., "Sensorless speed measurement using harmonic spectral estimation in induction machine drives," *IEEE Transactions on Power Electronics*, vol. 11, no. 1, pp. 66–73, 1996.

Hurst, K. D., Habetler, T. G., Griva, G., and Profumo, F., "Zero-speed tacholess IM torque control: Simply a matter of stator voltage integration," *IEEE Transactions on Industry Applications*, vol. 34, no. 4, pp. 790–795, 1998.

Hurst, K. D., Habetler, T. G., Griva, G., Profumo, F., and Jansen, P. L., "A self-tuning closed-loop flux observer for sensorless torque control of standard induction machines," *IEEE Transactions on Power Electronics,* vol. 12, no. 5, pp. 807–815, 1997.

Ishida, M., and Iwata, K., "Steady-state characteristics of a torque and speed control system of an induction motor utilizing rotor slot harmonics for a slip frequency sensing," *IEEE Transactions on Power Electronics*, vol. 2, no. 3, pp. 257–263, 1987.

Jang, J. -S. R., and Sun, C. -T., "Neuro-fuzzy modeling and control," *Proceedings of the IEEE*, vol. 83, no. 3, pp. 378–406, 1995.

Jansen, P. L., and Lorenz, R. D., "Transducerless position and velocity estimation in induction and salient a.c. machines," *IEEE Transactions on Industry Applications*, vol. 31, no. 2, pp. 240–247, 1995.

Jansen, P. L., and Lorenz, R. D., "Transducerless field orientation concepts employing saturation-induced saliences in induction motors," *IEEE Transactions on Industry Applications*, vol. 32, no. 6, pp. 1380–1393, 1996.

Jansen, P. L., Lorenz, R. D., and Novotny, D. W., "Observer-based direct field orientation: Analysis and comparison of alternative methods," *IEEE Transactions on Industry Applications*, vol. 30, no. 4, pp. 945–953, 1994.

Jezernik, K., Curk, B., and Harnik, J., "Variable structure field oriented control of an induction motor drive," *Proceedings of EPE'91*, pp. 161–166, 1991.

Jiang, J., and Holtz, J., "High dynamic speed sensorless ac drive with on-line model parameter tuning for steady-state accuracy," *IEEE Transactions on Industrial Electronics*, vol. 44, no. 2, pp. 240–246, 1997.

Kang, J. K., and Sul, S. K., "New direct torque control of induction motor for minimum torque ripple and constant switching frequency," *IEEE Transactions on Industry Applications*, vol. 35, no. 5, pp. 1076–1082, 1999.

Kanmachi, T., and Takahashi, I., "Sensorless speed control of an induction motor with no influence of secondary resistance variation," *Conference Record of IEEE-IAS'93*, pp. 408–413, 1993.

Kazmierkowski, M. P., and Kopcke, H. J., "A simple control system for current source inverter-fed induction motor drives," *IEEE Transactions on Industry Applications*, vol. 21, no. 3, pp. 617–623, 1985.

Kazmierkowski, M. P., and Sulkowski, W., "A novel vector control scheme for transistor PWM inverter-fed induction motor drives," *IEEE Transactions on Industrial Electronics*, vol. 38, no. 1, pp. 41–47, 1991.

Kerkman, R. J., Leggate, D., and Skibinski, G. L., "Interaction of drive modulation and cable parameters on ac motor transients," *IEEE Transactions on Industry Applications*, vol. 33, no. 3, pp. 722–731, 1997.

Kerkman, R. J., Seibel, B. J., Rowan, T. M., and Schlegel, D. W., "A new flux and stator resistance identifier for a.c. drives," *IEEE Transactions on Industry Applications*, vol. 32, no. 3, pp. 585–593, 1996.

Kerkman, R. J., Thunes, J. D., Rowan, T. W., and Schlegel, D. W., "A frequency based determination of transient inductance and rotor resistance for field commissioning purposes," *IEEE Transactions on Industry Applications*, vol. 32, no. 3, pp. 577–584, 1996.

Khambadkone, A. M., and Holtz, J., "Vector controlled induction motor drive with a self-commissioning scheme," *IEEE Transactions on Industrial Electronics*, vol. 38, no. 5, pp. 322–327, 1991.

Kim, H. W., and Sul, S. K., "A new motor speed estimator using Kalman filter in low-speed range," *IEEE Transactions on Industrial Electronics*, vol. 43, no. 4, pp. 498–504, 1996.

Kim, K. C., Ortega, R., Charara, A., and Vilain, J. P., "Theoretical and experimental comparison of two nonlinear controllers for current-fed induction motors," *IEEE Transactions on Control Systems Technology*, vol. 5, no. 3, pp. 338–348, 1997.

Kim, S. H., and Sul, S. K., "Maximum torque control of an induction machine in the field-weakening region," *IEEE Transactions on Industry Applications*, vol. 31, no. 4, pp. 787–794, 1995.

Kim, S. H., and Sul, S. K., "Voltage control strategy for maximum torque operation of an induction machine in the field-weakening region," *IEEE Transactions on Industrial Electronics*, vol. 44, no. 4, pp. 512–517, 1997.

Kim, Y. R., Sul, S. K., and Park, M. H., "Speed sensorless vector control of induction motor using an extended Kalman filter," *IEEE Transactions on Industry Applications*, vol. 30 no. 5, pp. 1225–1233, 1994.

Kioskeridis, I., and Margaris, N., "Loss minimization in induction motor adjustable-speed drives," *IEEE Transactions on Industrial Electronics*, vol. 43, no. 1, p. 226–231, 1996.

Kirshen, D. S., Novotny, D. W., and Lipo, T. A., "On line efficiency optimization of a variable frequency induction motor drive," *IEEE Transactions on Industry Applications*, vol. 21, no. 4, pp. 610–615, 1985.

Koga, K., Ueda, R., and Soneda, T., "Constitution of V/f control for reducing the steady-state error to zero in induction motor drive system," *IEEE Transactions on Industry Applications*, vol. 28, no. 2, pp. 463–471, 1992.

Kreindler, L., Moreira, J. C., Testa, A., and Lipo, T. A., "Direct field orientation controller using the stator phase voltage third harmonics," *IEEE Transactions on Industry Applications*, vol. 30, no. 2, pp. 441–447, 1994.

Krishnan, R., and Bharadwaj, A. S., "A review of parameter sensitivity and adaptation in indirect vector controlled induction motor drive system," *IEEE Transactions on Power Electronics*, vol. 6, no. 4, pp. 695–703, 1991.

Kubota, H., and Matsuse, K., "Speed sensorless, field oriented control of induction motor with rotor resistance adaptation," *IEEE Transactions on Industry Applications*, vol. 30, no. 5, pp. 1219–1224, 1994.

Kubota, H., Matsuse, K., and Nakano, T., "DSP-based speed adaptive flux observer of induction motor," *IEEE Transactions on Industry Applications*, vol. 29, no. 2, pp. 344–348, 1993.

Kung, Y. S., Liaw, C. M., and Ouyang, M. S., "Adaptive speed control for induction motor drives using neural networks" *IEEE Transactions on Industrial Electronics*, vol. 42, no. 1, pp. 25–32, 1995.

Lai, M. F., Nakano, M., and Hsieh, G. C., "Application of fuzzy logic in the phase-locked loop speed control of induction motor drive," *IEEE Transactions on Industrial Electronics*, vol. 43, no. 6, pp. 630–639, 1996.

Lascu, C., Boldea, I., and Blaabjerg, F., "A modified direct torque control for induction motor sensorless drive," *IEEE Transactions on Industry Applications*, vol. 36, no. 1, pp. 122–130, 2000.

Lee, J. S., Takeshita, T., and Matsui, N., "Stator-flux-oriented sensorless induction motor drive for optimum low-speed performance," *IEEE Transactions on Industry Applications*, vol. 33, no. 5, pp. 1170–1176, 1997.

Levi, E., "Rotor flux oriented control of induction machines considering the core loss," *Electric Machines and Power Systems*, vol. 24, no. 1, pp. 37–50, 1996.

Levi, E., "Impact of cross-saturation on accuracy of saturated induction machine models," *IEEE Transactions on Energy Conversion*, vol. 12, no. 3, pp. 211–216, 1997.

Lima, A. M. N., Jacobina, C. B., and de Souza, E. B., "Nonlinear parameter estimation of steady-state induction motor models," *IEEE Transactions on Industrial Electronics*, vol. 44, no. 3, pp. 390–397, 1997.

Lorenz, R. D., and Lawson, D. B., "A simplified approach to continuous on-line tuning of field oriented induction motor drives," *IEEE Transactions on Industry Applications*, vol. 26, no. 3, pp. 420–424, 1990.

Lorenz, R. D., Lipo, T. A., and Novotny, D. W., "Motion control with induction motors," *Proceedings of the IEEE*, vol. 82, no. 8, pp. 1215–1240, 1994.

Lorenz, R. D., and Van Patten, K. W., "High-resolution velocity estimation for all-digital, ac servo drives," *IEEE Transactions on Industry Applications*, vol. 27, no. 4, pp. 701–705, 1991.

Lorenz, R. D., and Yang, S. M., "Efficiency-optimized flux trajectories for closed-cycle operation of field-oriented induction machine drives," *IEEE Transactions on Industry Applications*, vol. 28, no. 3, pp. 574–579, 1992.

Manz, L., "Applying adjustable-speed drives to three-phase induction NEMA frame motors," *IEEE Transactions on Industry Applications*, vol. 33, no. 2, pp. 402–407, 1997.

Marchesoni, M., Segarich, P., and Soressi, E., "A simple approach to flux and speed observation in induction motor drives," *IEEE Transactions on Industrial Electronics*, vol. 44, no. 4, pp. 528–535, 1997.

Margaris, N., and Kioskeridis, I., "A method for the estimation of the rotor time constant in the indirect vector controlled drive," *IEEE Transactions on Industrial Electronics*, vol. 43, no. 1, pp. 232–233, 1996.

Mir, S., Elbuluk, M. E., and Zinger, D. S., "PI and fuzzy estimators for tuning the stator resistance in direct torque control of induction machines," *IEEE Transactions on Power Electronics*, vol. 13, no. 2, pp. 279–287, 1998.

Miyashita, I., Imayanagida, A., and Koga, T., "Recent industrial application of speed sensorless vector control in Japan," *Proceedings IECON'94*, pp. 1573–1578, 1994.

Miyashita, I., and Ohmori, Y., "A new speed observer for induction motor using the speed estimation technique," *Proceedings EPE'93*, pp. 349–353, 1993.

Moreira, J. C., Hung, K. T., Lipo, T. A., and Lorenz, R. D., "A simple and robust adaptive controller for detuning correction in field oriented induction machines," *IEEE Transactions on Industry Applications*, vol. 28, no. 6, pp. 1359–1366, 1992.

Moreira J. C., Lipo, T. A., and Blasko, V., "Simple efficiency maximizer for an adjustable frequency induction motor drive," *IEEE Transactions on Industry Applications*, vol. 27, no. 5, pp. 940–946, 1991.

Munoz-Garcia, A., Lipo, T. A., and Novotny, D. W., "A new induction motor V/f control method capable of high-performance regulation at low speeds," *IEEE Transactions on Industry Applications*, vol. 34, no. 4, pp. 813–821, 1997.

Nash, J. M., "Direct torque control, induction motor vector control without an encoder," *IEEE Transactions on Industry Applications*, vol. 33, no. 2, pp. 333–341, 1997.

Nicklasson, P. J., Ortega, R., and Espinosa-Perez, G., "Passivity-based control of a class of Blondel-Park transformable electric machines," *IEEE Transactions on Automatic Control*, vol. 42, no. 5, pp. 629–647, 1997.

Nilsen, R., and Kazmierkowski, M. P., "Reduced-order observer with parameter adaptation for fast rotor flux estimation in induction machine," *IEE Proceedings*, vol. 136, pt. D, no. 1, pp. 35–43, 1989.

Noguchi, T., Kondo, S., and Takahashi, I., "Field-oriented control of an induction motor with robust on-line tuning of its parameters," *IEEE Transactions on Industry Applications*, vol. 33, no. 1, pp. 35–42, 1997.

Noguchi, T., Yamamoto, M., Kondo, S., and Takahashi, I., "Enlarging switching frequency in direct torque-controlled inverter by means of dithering," *IEEE Transactions on Industry Applications*, vol. 35, no. 6, pp. 1358–1366, 1999.

Nordin, K. B., Novotny, D. W., and Zinger, D. S., "The influence of motor parameter deviations in feedforward field orientation drive systems," *IEEE Transactions on Industry Applications*, vol. 21, no. 4, pp. 1009–1015, 1985.

Ohnishi, K., Matsui, N., and Hori, Y., "Estimation, identification, and sensorless control in motion control systems," *Proceedings of the IEEE,* vol. 82, no. 8, pp. 1253–1265, 1994.

Ohtani, T., Takada, N., and Tanaka, K., "Vector control of induction motors without shaft encoder," *IEEE Transactions on Industry Applications,* vol. 28, no. 1, pp. 157–164, 1992.

Orlowska-Kowalska, T., "Application of extended Luenberger observer for flux and rotor time-constant estimation in induction motor drives," *IEE Proceedings,* vol. 136, pt. D, no. 6, pp. 324–330, 1989.

Ortega, R., Per, J., and Espinoza-Perez, G., "On speed control of induction motors," *Automatica,* vol. 32, no. 3, pp. 455–460, 1996.

Osama, M., and Lipo, T. A., "Modeling and analysis of a wide-speed-range induction motor drive based on electronic pole changing," *IEEE Transactions on Industry Applications,* vol. 33, no. 5, pp. 1177–1184, 1997.

Peng, F. Z., Fukao, T., and Lai, J. S., "Robust speed identification for speed-sensorless vector control of induction motors," *IEEE Transactions on Industry Applications,* vol. 30, no. 5, pp. 1234–1240, 1994.

Profumo, F., Griva, G., Pastorelli, M., Moreira, J., and De Doncker, R., "Universal field oriented controller based on air gap flux sensing via third harmonic stator voltage," *IEEE Transactions on Industry Applications,* vol. 30, no. 2, pp. 448–455, 1994.

Ran, L., Gokani, S., Clare, J., Bradley, K. J., and Christopoulos, C., "Conducted electromagnetic emissions in induction motor drive systems," Part 1 and 2, *IEEE Transactions on Power Electronics,* vol. 13, no. 4, pp. 757–776, 1998.

Rendusara, D. A., and Enjeti, P. N., "An improved inverter output filter configuration reduces common and differential modes dv/dt at the motor terminals in PWM drive system," *IEEE Transactions on Power Electronics,* vol. 13, no. 6, pp. 1135–1143, 1998.

Ribeiro, L. A., Jacobina, C. B., and Lima, A. M. N., "Linear parameter estimation for induction machines considering the operating conditions," *IEEE Transactions on Power Electronics,* vol. 14, no. 1, pp. 62–73, 1999.

Ribeiro, L. A., Jacobina, C. B., Lima, A. M. N., and Oliveira, A. C., "Parameter sensitivity of MRAC models employed in IFO-controlled ac motor drive," *IEEE Transactions on Industrial Electronics,* vol. 44, no. 4, pp. 536–545, 1997.

Rowan, T. M., Kerkman, R. J., and Leggate, D., "A simple on-line adaption for indirect field orientation of an induction machine," *IEEE Transactions on Industry Applications,* vol. 27, no. 4, pp. 720–727, 1991.

Rubin, N. P., Harley, R. G., and Diana, G., "Evaluation of various slip estimation techniques for an induction machine operating under field-oriented control conditions," *IEEE Transactions on Industry Applications*, vol. 28, no. 6, pp. 1367–1375, 1992.

Sabanovic, A., and Bilalovic, F., "Sliding mode control of ac drives," *IEEE Transactions on Industry Applications*, vol. 25, no. 1, pp. 70–74, 1989.

Schauder, C. D., "Adaptive speed identification for vector control of induction motors without rotational transducers, *IEEE Transactions on Industry Applications*, vol. 28, no. 5, pp. 1054–1061, 1992.

Schroedl, M., and Wieser, R. S., "EMF-based rotor flux detection in induction motors using virtual short circuits," *IEEE Transactions on Industry Applications*, vol. 34, no. 1, pp. 142–147, 1997.

Seibel, B. J., Rowan, T. M., and Kerkman, R. J., "Field-oriented control of an induction machine in the field-weakening region with dc-link and load disturbance rejection," *IEEE Transactions on Industry Applications*, vol. 33, no. 6, pp. 1578–1584, 1997.

Seok, J. K., and Sul, S. K., "Optimal flux selection of an induction machine for maximum torque operation in flux-weakening region," *IEEE Transactions on Power Electronics,* vol. 14, no. 4, pp. 700–708, 1999.

Shyu, K. K., and Shieh, H. J., "A new switching surface sliding-mode speed control for induction motor drive systems," *IEEE Transactions on Power Electronics,* vol. 11, no. 4, pp. 660–667, 1996.

Simoes, G., and Bose, B. K., "Neural network based estimation of feedback signals for a vector controlled induction motor drive," *IEEE Transactions on Industry Applications*, vol. 31, no. 3, pp. 620–629, 1995.

Skibinski, G. L., Kerkman, R. J., and Schlegel, D., "EMI emissions of modern PWM ac drives," *IEEE Industry Applications Magazine*, vol. 5, no. 6, pp. 47–81, 1999.

Slemon, G. R., "Modelling of induction machines for electric drives," *IEEE Transactions on Industry Applications*, vol. 25, no. 6, pp. 1126–1131, 1989.

Sng, E. K. K., Liew, A. C., and Lipo, T. A., "New observer-based DFO scheme for speed sensorless field-oriented drives for low-zero-speed operation," *IEEE Transactions on Power Electronics,* vol. 13, no. 5, pp. 959–968, 1998.

Spiteri Staines, C., Asher, G. M., and Bradley, K. J., "A periodic burst injection method for deriving rotor position in saturated cage-salient induction motors without a shaft encoder," *IEEE Transactions on Industry Applications*, vol. 35, no. 4, pp. 851–858, 1999.

Stephan, J., Bodson, M., and Chiasson, J., "Real-time estimation of the parameters and fluxes of induction motors," *IEEE Transactions on Industry Applications*, vol. 30, no. 3, pp. 746–759, 1994.

Sullivan, C. R., Kao, C., Acker, B. M., and Sanders, S. R., "Control systems for induction machines with magnetic saturation," *IEEE Transactions on Industrial Electronics*, vol. 43, no. 1, pp. 142–152, 1996.

Sullivan, C. R., and Sanders, S. R., "Models for induction machines with magnetic saturation in the main flux path," *Conference Record of IEEE-IAS'92*, pp. 123–131, 1992.

Tadakuma, S., Tanaka, S., Naitoh, H., and Shimane, K., "Improvement of robustness of vector-controlled induction motors using feedforward and feedback control," *IEEE Transactions on Power Electronics,* vol. 12, no. 1, pp. 221–227, 1997.

Tajima, H., and Hori, Y., "Speed sensorless field-orientation control of the induction machine," *IEEE Transactions on Industry Applications*, vol. 29, no. 1, pp. 175–180, 1993.

Takahashi, I., and Noguchi, T., "A new quick-response and high-efficiency strategy of an induction motor," *IEEE Transactions on Industry Applications*, vol. 22, no. 7, pp. 820–827, 1986.

Takahashi, I., and Ohmori, Y., "High performance direct torque control of an induction motor," *IEEE Transactions on Industry Applications*, vol. 25, no. 2, pp.257–264, 1989.

Takano, A., "Quick-response torque-controlled induction motor drives using phase-locked loop speed control with disturbance compensation," *IEEE Transactions on Industrial Electronics,* vol. 43, no. 6, pp. 640–646, 1996.

Toliyat, H. A., Arefeen, M. S., Rahman, K. M., and Figoli, D., "Rotor time constant updating scheme for a rotor flux-oriented induction motor drive," *IEEE Transactions on Power Electronics,* vol. 14, no. 5, pp. 850–857, 1999.

Trzynadlowski, A. M., "Safe operating and safe design areas of induction motor drives," *IEEE Transactions on Industry Applications*, vol. 31 no. 5, pp. 1121–1128, 1995.

Trzynadlowski, A. M., Zigliotto, M., and Bech, M. M., "Random pulse width modulation quiets motors, reduces EMI," *PCIM Power Electronics Systems Magazine*, pp. 55–58, February 1999.

Utkin, V. I., "Sliding mode control design principles and applications to electric drives," *IEEE Transactions on Industrial Electronics*, vol. 40, no. 1, pp. 23–26, 1993.

Vagati, A., Fratta, A., Franceschini, G., and Rosso, P., "AC motors for high-performance drives: A design-based comparison," *IEEE Transactions on Industry Applications*, vol. 32 no. 5, pp. 1211–1219, 1996.

Verghese, G. C., and Sanders, S. R., "Observers for faster flux estimation in induction machines," *IEEE Transactions on Industrial Electronics*, vol. 35, no. 1, pp. 85–94, 1988.

Von Jouanne, A., and Enjeti, P., "Design considerations for an inverter output filter to mitigate the effects of long motor leads in ASD applications," *IEEE Transactions on Industry Applications*, vol. 33, no. 5, pp. 1138–1145, 1997.

Von Jouanne, A., Spee, R., Faveluke, A., and Bhowmik, S., "Voltage sag ride-through for adjustable-speed drives with active rectifiers," *IEEE Transactions on Industry Applications*, vol. 34, no. 4, pp. 1270–1277, 1997.

Von Jouanne, A., Zhang, H., and Wallace, A. K., "An evaluation of mitigation techniques for bearing currents, EMI and overvoltages in ASD applications," *IEEE Transactions on Industry Applications*, vol. 34, no. 5, pp. 1113–1122, 1997.

Wade, S., Dunnigan, M. W., and Williams, B. W., "A new method of rotor resistance estimation for vector-controlled induction machines," *IEEE Transactions on Industrial Electronics*, vol. 44, no. 2, pp. 247–257, 1997.

Wang, M., Levi, E., and Jovanovic, M., "Compensation of parameter variation effects in sensorless indirect vector controlled induction machines using model-based approach," *Electric Machines and Power Systems*, vol. 27, no. 4, pp. 1001–1027, 1999.

Wu, X., Q., and Steimel, A., "Direct self control of induction machines fed by a double three-level inverter," *IEEE Transactions on Industrial Electronics*, vol. 44, no. 4, pp. 519–527, 1997.

Xu, X., and Novotny, D. W., "Implementation of direct stator flux orientation control on a versatile DSP based system," *IEEE Transactions on Industry Applications*, vol. 27, no. 4, pp. 694–700, 1991.

Yang, G., and Chin, T. H., "Adaptive speed identification scheme for vector controlled speed sensorless inverter induction motor drives," *IEEE Transactions on Industry Applications*, vol. 29, no. 4, pp. 820–825, 1993.

Yang, J. H., Yu, W. H., and Fu, L. C., "Nonlinear observer-based adaptive tracking control for induction motors with unknown load," *IEEE Transactions on Industrial Electronics*, vol. 42, no. 6, pp. 579–586, 1995.

Yong, S. H., Choi, J. W., and Sul, S. K., "Sensorless vector control of induction machine using high frequency current injection," *Conference Record of IEEE-IAS'94*, pp. 503–508, 1994.

Zhang, J. and Barton, T. H., "A fast variable structure current controller for an induction motor drive," *IEEE Transactions on Industry Applications*, vol. 26, no. 3, pp. 415–419, 1990.

Zhen, L., and Xu, L., "On-line fuzzy tuning of indirect field-oriented induction machine drives," *IEEE Transactions on Power Electronics*, vol. 13, no. 1, pp. 134–141, 1998.

GLOSSARY OF SYMBOLS

PRINCIPAL SYMBOLS

a, b, c	switching variables
d	duty ratio
E	energy, J
e	space vector of EMF, V
e	instantaneous EMF, V
$\mathscr{F}$	space vector of magnetomotive force, A
f	frequency, Hz
$\hat{I}$	phasor of current, A
I	constant value of current (e.g., rms or peak), A
i	space vector of current, A
i	instantaneous current, A
J	mass moment of inertia, kgm^2
L	inductance, H
m	modulation index, or mass, kg
n	rotational speed, r/min
P	real power, W
p	differentiation operator (d/dt), s^{-1}

p_p	number of pole pairs
R	resistance, Ω
S	apparent power, VA
s	slip
T	torque, Nm, or period, s
t	time, s
u	linear speed, m/s
$\hat{V}$	phasor of voltage, V
V	constant value of voltage (e.g., rms or peak), V
v	space vector of voltage, V
v	instantaneous voltage, V
X	reactance, Ω
α, β, φ	angle, rad
η	efficiency
Θ	angular position, rad
$\hat{\Lambda}$	phasor of flux, Wb
Λ	rms value of flux, Wb
λ	space vector of flux, Wb
σ	leakage factor
τ	time constant, s
ω	angular velocity or radian frequency, rad/s

SUBSCRIPTS

A, B, C	phase A, phase B, phase C
a, b, c	phase A, phase B, phase C
act	actual
br	braking
c	capacitive
cm	common mode
cr	critical
D	D-axis of revolving reference frame
d	d-axis of stator reference frame
dc	direct current
elec	electrical
eq	equivalent
f	field or flux
h	high frequency
i, in	input
L	line or load
l, σ	leakage
M	motor
m	magnetizing or peak

max	maximum
mech	mechanical
N	neutral
o	rotor of two-pole equivalent motor
out	output
Q	Q-axis of revolving reference frame
q	q-axis of stator reference frame
R, r	rotor
S, s	stator
rat	rated
sh	slot harmonic
sl	slip
smp	sampling
st	starting
sw	switching
syn	synchronous

SUPERSCRIPTS

e	revolving reference frame
s	stator reference frame
T	transposed
*	conjugate or reference
'	estimated

ACRONYMS

ANFIS	Adaptive NeuroFuzzy Inference System
ARCP	Auxiliary Resonant Commutated Pole
CSI	Current Source Inverter
CVH	Constant Volts/Hertz
DFO	Direct Field Orientation
DSC	Direct Self-Control
DSP	Digital Signal Processor
DTC	Direct Torque Control
EDM	Electric Discharge Machining
EMF	Electromotive Force
EMI	Electromagnetic Interference
ESO	Essential Service Override
FFT	Fast Fourier Transform
FO	Field-Oriented

GTO	Gate Turn-Off Thyristor
IFO	Indirect Field Orientation
IGBT	Insulated Gate Bipolar Transistor
ITR	Ideal Transformer
MI	Machine Intelligence
MMF	Magnetomotive Force
MRAS	Model Reference Adaptive System
PF	Power Factor
PI	Proportional-Integral
PLL	Phase-Locked Loop
PWM	Pulse Width Modulation
RDCL	Resonant DC Link
RF	Radio Frequency
SCR	Silicon Controlled Rectifier
SV	Sensorless Vector
SVPWM	Space Vector Pulse Width Modulation
THD	Total Harmonic Distortion
VSC	Variable Structure Control
VSI	Voltage Source Inverter
ZCS	Zero-Current Switching
ZVS	Zero-Voltage Switching

INDEX

Ac voltage controllers, 56
Abnormal operating conditions, 52–54
Adaptive neurofuzzy inference system (ANFIS), 170–173
Adjustable-speed drives (ASDs), 2, 3
  control of, 4
Air-gap flux, 26, 110
  orientation, 129–134
Armature reaction, 120
Autotransformer starting, 44–45
Auxiliary resonant commutated pole (ARCP) inverter, 68–69

Back propagation algorithm, 165
Bang-bang control, 83–84
Bearing failures, 53
Bias weight, 165
Blanking time, 65
Blocked-rotor test, 192
Braking, 35–36
  dc-current, 198
  dynamic, 47, 48–51, 198

electric, 11, 198
flux, 198
resistor, 70–71

Cascade systems, 2
Centroid technique, 168
Chattering, 164
Closed-loop scalar speed control, 101–102
Commissioning, parameter adaptation and self-, 191–197
Common-mode voltage, 89–90
Constant power area, 99
Constant power connection, 52
Constant torque area, 99
Constant torque characteristic, 7
Constant torque connection, 52
Constant Volts/Hertz (CVH) method, 97–100, 198
Construction of induction motors, 15–17
Cooling fan/fins, 16
Corner folding, 149
Critical drives, 198
Critical slip, 30

Current equation, 184
Current source inverters (CSIs), 66
  control of, 85–88
  field orientation, 134–135
Current source PWM rectifiers, 63–64
Cycloconverters, 56

Dc-current braking, 198
Dc link, 4
Dc motor, torque production and control in, 119–121
Dead time, 65, 198
Defuzzifier, 168
Developed torque, 27–31, 113–114
Direct field orientation (DFO), 124–126
Direct self-control (DSC), 148–155
Direct torque control (DTC), 140–148
Drive systems, 3–4
Duty ratio, 77–78
Dynamic braking, 47, 48–51, 198
Dynamic model
  equations of the induction motor, 111–114
  revolving reference frame, 114–117
  space vectors of motor variables, 107–111
Dynamic torque, 5

Electrical braking, 11, 198
Electrical discharge machining (EDM), 89
Electric drive system, 3
Electromagnetic interference (EMI), 4, 68
  conducted, 88–89
Equations of the induction motor, 111–114
Equivalent wheel, 5–6
Essential Service Override (ESO), 200
Estimators, open-loop, 126, 179
  flux calculators, 179–183
  speed calculators, 183–191
Extended Kalman Filter (EKF), 189–191

Faraday's law, 23
Fast Fourier Transform (FFT), 191
Feedforward neural network, 164–166
Field orientation
  development of, 119
  direct, 124–126
  drives with current source inverters, 134–135, 198, 199
  indirect, 126–129
  principles of, 121–123
  stator- and air-gap flux, 129–134
Field weakening mode, 99
Firing strength, 168
Flashover, 89

Flux braking, 198
Flux calculators, 179–183
Flux
  airgap, 26
  leakage, 26
  rotor, 26
  stator, 26
  virtual, 149
Flux-linkage vectors, 110
Flux-producing current, 103–104
Flux vectors, 110, 199
Four-pole stator, 20–21
Frequency changers, 69–71
Fuzzifier, 167
Fuzzy controllers, 166–170
Fuzzy logic, 166

Gamma (Γ) model of the induction motor, 94
Generating mode, 35, 36
GTOs (gate turn-off thyristors), 69, 134

Hall sensors, 124
Hard switching, 67–68
Harmonic traps, 61, 88
High-efficiency motors, 2
Hysteresis controllers, 83–84, 149, 163

IGBTs (insulated-gate bipolar transistors), 69, 198
  non-punch-through, 61
Impedance starting, 45
Indirect field orientation (IFO), 126–129
Induction generators, 36–40
Induction motors
  construction of, 15–17
  operating principles of, 1–3
  variables controlled in, 159–161
Inference engine, 167–168
Input power, 31, 34
Inverse-gamma (Γ') model of the induction motor, 95–96
Inverter mode, 59
Inverters, 3–4, 56, 64–69
  control of current source, 85–88
  control of voltage source, 71–85

Kalman filters, 188–189
  extended, 189–191
Kinetic energy, 4–5

Left-hand rule, 23
  Life span of motors, 2–3
Linear speed controllers, 162–163

Loads
  common, 4–9
  nonlinear, 60
Luenberger state observer, 187–188

Machine intelligence (MI) controllers, 164–173
Magnetic field, revolving, 17–23
Magnetomotive forces (MMFs), 19, 108–109
MATLAB, 34
MCTs (MOSFET-controlled thyristors), 69
Mechanical characteristic, 6–7
Membership function, 166
Membership grade, 166
Model Reference Adaptive System (MRAS),
  183–187
Modulation index, 78
Motoring, 35
Multilevel voltage source inverters, 66–67

Neurofuzzy controllers, 170–173
Neutral-clamped inverter, 67
Newton's second law, 5
Noise, tonal, 90–91
No-load test, 192
Nonlinear loads, 60

Observers, closed-loop, 126, 179
  flux calculators, 179–183
  speed calculators, 183–191
Open-frame motors, 16
Open-loop scalar speed control, 97–100
Operating areas, 7
Operating quadrants, 10–11
Overload, mechanical, 53–54
Overvoltages, 89

Parameter adaptation, 191–197
Partly enclosed motors, 16
Passive filters, 60–61
Phase-lock loop (PLL), 163
Phase loss, 53
Plugging, 47–48
Pole changing, 3, 51–52
Position control, 173–175
Power electronic converters
  control of current source inverters, 85–88
  control of stator voltage, 55–56
  control of voltage source inverters, 71–85
  frequency changers, 69–71
  inverters, 56, 64–69
  rectifiers, 56–64
  side effects, 88–91
Power factor (PF), 31, 34

Prime mover, 10
Progressive-torque characteristic, 7
Proportional-integral (PI) adaptive scheme, 188
Pull-out torque, 30
Pulse width modulation (PWM), 61
  inverters, 73–80, 87
  rectifiers, 61–64

Rated power, 31, 32
Rated speed, 31, 32
Rated stator voltage, 32, 33–34
Rated torque, 31, 32
Reactance, stator and rotor, 28–29
Rectifiers, 4, 56–64
Regressive-torque characteristic, 7
Resistive-capacitive (RC) filters, 89
Resonant dc link (RDCL) inverter, 68
Reversing, 51
Revolving reference frame, 114–117
Right-hand rule, 22–23
Rotor, 16
  angular velocity, 39, 95
  bars, cracked, 53
  current vector, 110, 113
  equation, 184
  flux, 26, 110, 113
  frequency, 94
  inductance, 112
  leakage factor, 133
  reactance, 28
  resistance, 24
  voltage vector, 113

Scalar control methods, 11–12
  closed-loop speed control, 101–102
  open-loop speed control, 97–100
  torque control, 102–105
  two-inductance equivalent circuits, 93–97
Self-commissioning and parameter adaptation,
  191–199
Sensorless drives, 198–199
  basic issues, 177–179
  commercial, 197–200
  flux calculators for, 179–183
  parameter adaptation and self-commissioning,
  191–197
  speed calculators, 183–191
Short circuits, 53
Shot-through, 65
Silicon controlled rectifiers (SCRs), 58–59
Six-step mode, 72–73
Sliding mode, 163
Slip, 23

critical, 30
 velocity, 23
Soft starting, 46–47
Soft-switching inverters, 68
Space-vector direct torque and flux control, 155–157
Space Vector Pulse Width Modulation (SVPWM), 77
Space vectors of motor variables, 107–111
Speed calculators, 183–191
Speed control, 161–164
Square-law torque connection, 52
Square-wave mode, 65, 72–73
Squashing function, 165
Squirrel-cage motors, 1, 2
 description of, 16–17
Starting, assisted, 44–47
Starting torque, 30
State duty ratios, 77–78
Static torque, 5, 6
Stator, 2
 core, 15
 currents, 18
 current vector, 109
 equation, 184
 flux, 26, 110, 113
 field orientation, 129–134
 inductance, 112
 reactance, 28
 resistance, 24
Stator voltage, 39
 control of, 55–56
 vector, 109
Steady-state equivalent circuit
 characteristics, 31–36
 description of, 24–27
 developed torque, 27–31
Switching frequency, 73
Switching function, 173–174
Switching intervals, 73
Switching variables, 71, 85–87
Synchronous speed, 22, 35
Synchronous velocity, 21

Thyristors, 58–59
T-model of the induction motor, 93–97
Torque
 angle, 138

braking, 49
driving, 40
dynamic, 5
production and control in the dc motor, 119–121
pull-out, 30
rated, 31, 32
scalar control of, 102–105
starting, 30
static, 5, 6
Torque-per-ampere ratio, 120
Torque-producing current, 103–104
Total harmonic distortion (THD), 150
Total leakage factor, 129
Totally enclosed motors, 16
Two-inductance equivalent circuits, 93–97
Two-pole stator, 20

Uncontrolled induction motors
 abnormal operating conditions, 52–54
 assisted starting, 44–47
 braking and reversing, 47–51
 operation of, 43–44
 pole changing, 51–52
User interface devices, 199

Variable structure control (VSC), 163
Vector control methods, 11–12
Virtual fluxes, 149
Voltage
 common-mode, 89–90
 control of stator, 55–56
 equation, 184
 noise, 88–89
 sags, 29, 44, 53
Voltage-current equation, 112–113
Voltage source inverters (VSIs), 64–65
 control of, 71–85
Voltage source PWM rectifiers, 61–63
Voltage space vectors, 74–75

Water-hammer effect, 51
Wound-rotor motors, 2
Wye-delta starting, 46
Wye-delta switch, 11

Zero-current switching (ZCS), 68
Zero-voltage switching (ZVS), 68

ˀnd bound by CPI Group (UK) Ltd, Croydon, CR0 4YY

12/05/2025

01866873-0001